Leckie×Leckie

Scotland's leading educational publishers

S1 to National 4
CHEMISTRY

PRACTICE QUESTION BOOK

Bob Wilson

S1 to N4 CHEMISTRY PRACTICE QUESTION BOOK

© 2018 Leckie & Leckie Ltd

00114062018

10 9 8 7 6 5 4 3 2 1

ISBN 9780008263645

Published by
Leckie & Leckie Ltd
An imprint of HarperCollins*Publishers*
Westerhill Road, Bishopbriggs, Glasgow, G64 2QT
T: 0844 576 8126   F: 0844 576 8131
leckieandleckie@harpercollins.co.uk   www.leckieandleckie.co.uk

Special thanks to
Jouve (layout and illustration); Ink Tank (cover design);
Project One Publishing Solutions (project management);
Jess White (proofread)

A CIP Catalogue record for this book is available from the British Library.

Acknowledgements
Cover image: © Howard Walker / Alamy Stock Photo

P1 © ANDREW LAMBERT PHOTOGRAPHY/SCIENCE PHOTO LIBRARY; P25 © Michael J Thompson / Shutterstock.com; P25 JeffreyRasmussen / Shutterstock.com; P26 © molekuul_be / Shutterstock.com; P27 © banluporn namnorin / Shutterstock.com; P43 © Michael J Thompson / Shutterstock.com; P48 © Shutterstock.com; P50 © Shutterstock.com; P52 © petarg / Shutterstock.com; P61 © Africa Studio / Shutterstock.com; P80 © Singkham / Shutterstock.com

Printed by CPI Group (UK) Ltd, Croydon CR0 4YY

ANSWERS

https://collins.co.uk/pages/scottish-curriculum-free-resources

Introduction

About this book

This book provides a resource to support you in your study of chemistry. This book follows the structure of the Leckie and Leckie *S1 to National 4 Chemistry Student Book*.

Questions have been written to cover the Key areas in the Units of National 3 Chemistry and National 4 Chemistry and the Chemistry Experiences and Outcomes (Es and Os) of Curriculum for Excellence (CfE) Science at 3rd and 4th levels.

Features

This exercise includes coverage of

Each chapter begins with references to the **N3** or **N4** key area and the curriculum level, **CL3** or **CL4**, and the Es and Os code it covers.

Exercise 6C Acidic gases and the environment

This exercise includes coverage of:
N3 Acids and bases
CL3 Planet Earth SCN 3-05b
CL3 Materials SCN 3-18a

Exercise questions

The exercises give a range of types of ques tions. There are questions practising the demonstration and application of knowledge, and questions practising skills of scientific inquiry.

 7 The table shows the average percentage of iron found in some ores of iron.

a Draw a bar chart of 'Iron ore' against 'Percentage of iron (%)'.

b Ores containing over 60% iron are often called usable ores.

  i Which of the ores in the table can be considered usable?

  ii Usable ores can be fed directly into a blast furnace to extract the iron. Blast furnaces operate at very high temperatures. Another substance has to be added to extract the iron.

  Suggest what the substance might be.

Iron ore	Percentage of iron (%)
magnetite	73
haematite	70
goethite	63
limonite	55
siderite	48

Hint

Where appropriate, **hints** are provided to give extra support.

Hint	Identify the elements and compounds in a word equation.
	Elements have their symbol as their formula, except for the diatomic elements, e.g. Cl_2.
	The formula of a compound can be worked out using valency.

Answers

Answers to all questions are provided online at:
https://collins.co.uk/pages/scottish-curriculum-free-resources

Skills of scientific inquiry

The table below shows the various skills of scientific inquiry practised in the book.

The following pages support the answering of skills questions. To help you with the skills questions it's best to have graph paper, a ruler, calculator and a sharp pencil. Highlighter pens can also be handy.

Skill	Meaning	Page
1 Selecting	Selecting information from graphs, charts and text	v
2 Presenting	Presenting information as graph and charts	vi
3 Processing	Processing information using calculations	vi
4 Planning	Planning or designing experiments including controls, variables and reliability	vii
5 Concluding	Drawing valid conclusions from experimental results	vii
6 Predicting	Making predictions based on evidence from experimental results	viii
7 Evaluating	Identifying strengths and suggesting improvements to weaknesses in experiments	viii

This section provides hints for the questions about skills of scientific inquiry listed above.
Follow the order of the steps in the diagrams: ① red, to ② green, to ③ blue, to ④ purple, to ⑤ orange.

Skill 1   Selecting information

You should be able to select relevant information from text, tables, charts, graphs and diagrams. The following example shows how to select information from a **diagram.**

The diagram below shows the results of a paper chromatography experiment carried out to separate and identify dyes in food colouring.

① Read the **stem** of the question carefully – it **tells you what the question is about** and contains vital information.

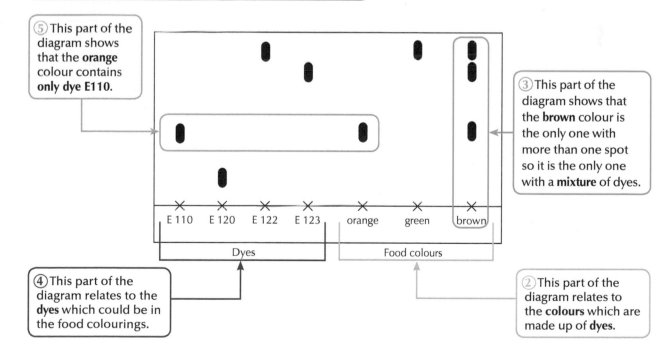

⑤ This part of the diagram shows that the **orange** colour contains **only dye E110.**

③ This part of the diagram shows that the **brown** colour is the only one with more than one spot so it is the only one with a **mixture** of dyes.

④ This part of the diagram relates to the **dyes** which could be in the food colourings.

② This part of the diagram relates to the **colours** which are made up of **dyes.**

| E 110 | E 120 | E 122 | E 123 | orange | green | brown |

Dyes   Food colours

Q a *Which food colour is a mixture of dyes?*

   b *Which food colour contains only dye E110?*

A a *Brown*

   b *Orange*

Skill 2 Presenting information

You should be able to **present** information by **completing** a line graph, bar chart or pie chart. The following example shows how to complete a **line graph**.

The progress of the reaction between marble chips and hydrochloric acid can be monitored by measuring the volume of gas produced over time. The results are shown in the table.

Q *Draw a line graph to show how the rate of reaction changes over time.*

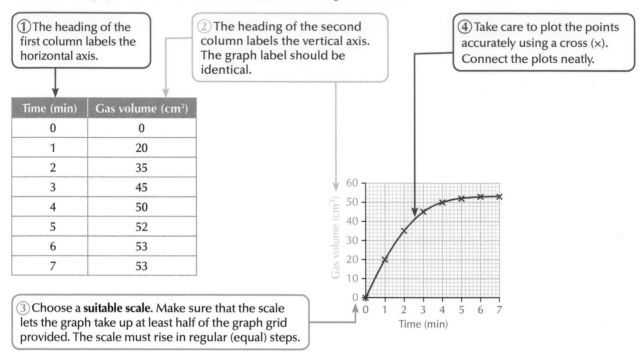

① The heading of the first column labels the horizontal axis.

② The heading of the second column labels the vertical axis. The graph label should be identical.

④ Take care to plot the points accurately using a cross (×). Connect the plots neatly.

Time (min)	Gas volume (cm³)
0	0
1	20
2	35
3	45
4	50
5	52
6	53
7	53

③ Choose a **suitable scale**. Make sure that the scale lets the graph take up at least half of the graph grid provided. The scale must rise in regular (equal) steps.

Skill 3 Processing

These examples show how to calculate **averages** and **percentages**.

Example 1

Groups of students carried out an experiment to see how many spoonfuls of sugar could dissolve in 100 cm³ of water. Their results are shown in the table below.

Q *Calculate the average number of spoonfuls of sugar which dissolve.*

Group	Number of spoonfuls
1	6
2	8
3	7
4	6
5	8

A $6 + 8 + 7 + 6 + 8 = 35$ ① Add all the results together.

$35 \div 5 = 7$ ② Divide the total by the number of groups.

Example 2

A 40 g rocksalt sample was found to contain 35.5 g of pure salt.

Q *Calculate the **percentage** of salt in the rocksalt sample.*

A $\dfrac{35.5}{40} = 0.8875$ ① Divide the mass of pure salt by the mass of rocksalt.

$0.8875 \times 100 = 88.75\%$ ② Multiply your answer by 100 to get the percentage.

Skill 4   Planning

You should be able to plan an experiment including the control of **variables** and ensuring **reliability**.

The diagram shows a method to investigate the effect of changing acid concentration (independent **variable**) on the rate of reaction (dependent **variable**) between zinc and hydrochloric acid.

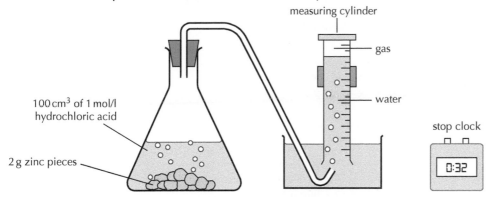

All other variables must be kept constant including temperature of the acid, mass and size of the zinc pieces and volume of acid.

Keeping all other variables constant ensures that any changes in the rate of reaction must be caused by the change in the concentration of the acid and not by other factors.

You will often be asked to **identify variables** which must be kept the same. Common variables include temperature or pH and masses, volumes and concentrations of substances in the experiment.

Experiments such as this are **repeated** so that **reliability** is improved. Reliability is a measure of the confidence in the results.

Sometimes, experiments have a built-in **control** which helps to make conclusions valid.

A **valid** conclusion is a fair conclusion which can be drawn from experimental results.

Skill 5   Concluding

You should be able to **draw conclusions** from experimental results.

The graph shows the results of the experiment which aimed to investigate how the rate of reaction between zinc and hydrochloric acid was affected by increasing the concentration of the acid.

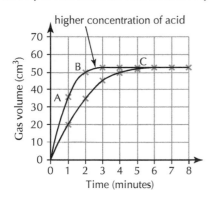

Q *Describe how the rate at which the gas is produced has been affected by increasing the concentration of the acid and draw a conclusion from this.*

A *Near the beginning of the reaction (A–B), the gas is produced faster with the higher concentration of acid. The final volume of gas produced (C) is the same in each experiment.*

Conclusion: The higher the concentration of acid, the faster the rate of reaction.

Skill 6 Predicting

You should be able to make a **prediction** from a set of experimental results. This involves thinking about what would happen if the experiment were tried in a different situation.

The graph shows the mass (weight) of potassium nitrate which dissolves in 100 cm³ of water at different temperatures up to 80 °C.

Q *Predict the mass of potassium nitrate which would dissolve at 100 °C.*

A *200 g per 100 cm³ water*

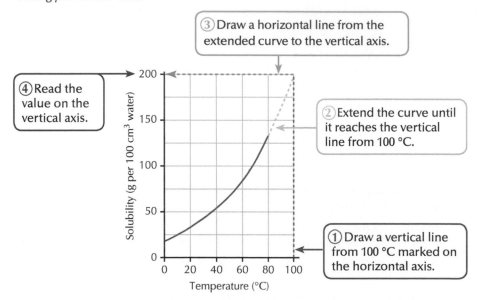

③ Draw a horizontal line from the extended curve to the vertical axis.

④ Read the value on the vertical axis.

② Extend the curve until it reaches the vertical line from 100 °C.

① Draw a vertical line from 100 °C marked on the horizontal axis.

Skill 7 Evaluating

You should be able to **evaluate** an experiment to identify strengths and weaknesses and suggest improvements.

The diagram shows apparatus designed to compare the energy produced when different foods are burned.

A burning spoon is filled with a food, set alight and held under the boiling tube which contains exactly 20 cm³ of water. The volume of water is measured in a measuring cylinder in each experiment. The temperature of the water is measured at the start and when the food stops burning. The thermometer can be read to 0.1 °C. The difference in temperature of the water gives an indication of the amount of energy produced by each food.

Q *Evaluate this experiment by identifying strengths and suggesting a method of improving weaknesses.*

A *Strengths*
The volume of water used in each experiment is the same and measured accurately using a measuring cylinder.
The temperature is measured to 0.1 °C accuracy.
Weaknesses
The same mass of food may not be used each time so it should be weighed out to improve the accuracy of the result.
The burning spoon may not be held at the same distance from the boiling tube each time if held in the hand. This could result in varying amounts of heat getting to the water. The burning spoon should be clamped at the same distance from the boiling tube each time.

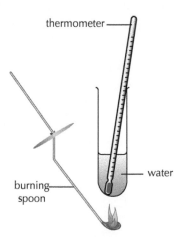

thermometer

water

burning spoon

1 Rates of reaction 1

Exercise 1A Chemical reactions

 State what always happens when a chemical reaction takes place.

 State **four** changes which could indicate that a chemical reaction has taken place.

 A student wrote a report about an experiment she carried out.

We added a piece of calcium to water. It started to fizz. After a while the calcium seemed to disappear. A milky mixture was left behind.

a Suggest what happened to the calcium, causing it to 'disappear'.

b Suggest what the 'fizz' is.

 The photograph shows what happens when a piece of magnesium metal is held in a Bunsen flame.

Give **two** pieces of evidence which suggest that a chemical reaction has taken place.

 Two colourless solutions were mixed together in a beaker. Although no change in appearance was seen, a chemical reaction had taken place.

a Suggest another change that might have taken place which would indicate that a reaction had taken place.

b Describe what could be done to show that the change in part **a** had actually taken place.

Exercise 1B Speeding up chemical reactions

1 **a** Fireworks contain gunpowder and metal compounds which produce different colours when they are heated.

   Explain why very finely ground gunpowder is used rather than lumps.

b Explain why milk stays fresher for longer when kept in a fridge.

c Hydrogen gas is produced when zinc reacts with hydrochloric acid.

   Suggest **two** things which could be done to the acid in order to speed up the rate at which hydrogen is produced.

2 Three different metals were added to separate test tubes containing hydrochloric acid.

A   B   C

a **i** In which test tube is the reaction fastest?

   ii Explain your answer to part **i**.

b Suggest **two** things you could do to the hydrochloric acid to speed up the reactions.

c Suggest **one** thing you could do to the metals to speed up the reactions.

> **Hint** Experimental conditions which can be changed, such as temperature and concentration, are known as **variables**. When one variable is changed all the others have to stay the same. This ensures that any difference observed in the rate of reaction is due only to the variable which has been changed.
>
> The **concentration** of a solution is a combination of the quantity of solute and volume of solvent it is dissolved in. The units of measurement can be grams per litre (g / l or g l^{-1}) or moles per litre (mol / l or mol l^{-1}). The important thing to remember is the number before the unit, e.g. a 0·5 mol l^{-1} solution is more concentrated than, for example, a 0·1 mol l^{-1} solution.

3 **a** State what a catalyst does in a chemical reaction.

b State why small pieces of catalyst work better than large lumps.

c Give **two** examples of catalysts being used in everyday life.

d A catalyst is unchanged at the end of a chemical reaction.

   Suggest why this is useful in the chemical industry.

4 Give **one** example of an everyday chemical reaction which happens:

a very fast

b very slowly.

5 A group of students carried out an experiment to see what effect changing the particle size of the reactants had on the rate of reaction. Here is the report written by one of the students.

We added small lumps of marble to hydrochloric acid. We saw bubbles of gas being produced quickly. We did the experiment again with large lumps of marble. Bubbles of gas were produced very slowly. When powdered marble was used the gas was produced fastest of all.

a State the aim of the experiment.

b State how the students could tell that a chemical reaction was taking place.

c Present the results the student obtained in a table. Use the headings 'Particle size' and 'Observations'.

d In each of the experiments state which conditions (variables) would have to be kept the same so that the comparisons were fair.

e What conclusion could the students draw from the results of their experiments?

6 Four experiments were carried out to compare the rate of reaction between chalk and hydrochloric acid under different conditions. The weight of chalk used in each experiment is the same.

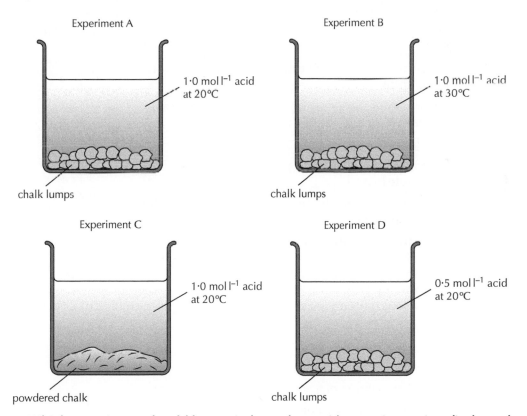

Experiment A — 1·0 mol l⁻¹ acid at 20°C — chalk lumps

Experiment B — 1·0 mol l⁻¹ acid at 30°C — chalk lumps

Experiment C — 1·0 mol l⁻¹ acid at 20°C — powdered chalk

Experiment D — 0·5 mol l⁻¹ acid at 20°C — chalk lumps

a Which experiment should be carried out along with experiment A to find out the effect of changing the particle size on the rate of reaction?

b i A student stated that experiments C and D should be carried out to compare the effect of changing the concentration of the acid.

State why this would not be a fair comparison.

ii State which two experiments should be carried out to compare the effect of changing the concentration of the acid.

c In which experiment would the reaction be slowest?

2 Rates of reaction 2

This chapter includes coverage of:

N4 Rates of reaction

Exercise 2A Monitoring the rate of a reaction

1 Part of the apparatus which can be used to study the rate at which excess hydrochloric acid reacts with lumps of chalk is shown. Carbon dioxide gas is produced. The gas is collected and the volume produced every minute measured.

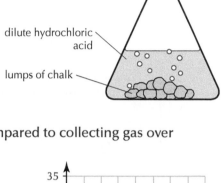

dilute hydrochloric acid

lumps of chalk

  a Make two copies of the apparatus. Add to the diagrams to show how the gas can be collected:

    i over water

    ii using a gas syringe.

    Label your diagrams.

  b Suggest one advantage using a gas syringe has compared to collecting gas over water.

  c A graph showing the volume of gas collected over time is shown.

    i At which part on the graph, A, B or C, is the rate of gas collection the highest?

    ii Explain your answer to part **i**.

    iii What volume of gas was produced after 1 minute?

    iv After how many minutes was the endpoint of the reaction reached?

  d Sketch the graph above and add another line to show what happens when the reaction is carried out with excess acid of a higher concentration. The mass and size of the lumps of chalk and the temperature of the reactants are the same as in the original experiment. (You do not need to use graph paper or include the scales on your sketch.)

2 The table shows the volume of gas produced when marble pieces react with hydrochloric acid.

  a Predict the volume of gas collected after 7 minutes.

  b Draw a graph of gas volume against time.

  c The experiment was repeated using powdered marble. All the other variables were kept the same.

    Add a line to the graph you drew in part **b** to show what happens when the reaction is carried out with powdered marble.

Time (min)	Gas volume (cm³)
0	0
1	25
2	45
3	60
4	70
5	75
6	75
7	

3 The rate at which magnesium ribbon reacts with sulfuric acid can be studied using the arrangement shown. Hydrogen gas is produced during the reaction.

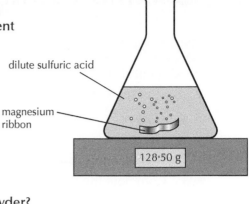

dilute sulfuric acid

magnesium ribbon

128·50 g

a Copy and complete the following sentence by choosing the correct word in the brackets.

As the reaction progresses the reading on the balance (**decreases** / **increases** / **stays the same**).

b How would the readings on the balance compare if the magnesium ribbon was replaced by the same mass of magnesium powder?

4 The graph shows the loss in mass over time when the same mass of zinc pieces is reacted with hydrochloric acid at 20 °C and 30 °C.

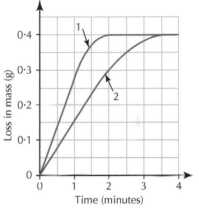

a i Which line, 1 or 2, is for the reaction carried out at 30 °C?

ii Explain your answer to part **i**.

b How many minutes does it take for the reaction represented by line 2 to reach its endpoint?

c i What is the total loss in mass in both experiments?

ii Suggest why the total loss in mass is the same in each experiment.

5 Chalk reacts with acid to produce carbon dioxide gas. The rate of the reaction can be followed by measuring the change in mass of the flask and its contents over time.

A graph of the results of such an experiment is shown.

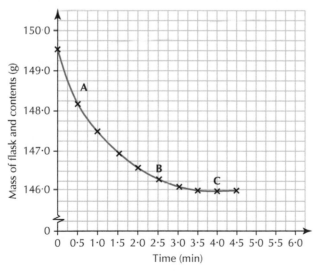

a At which point, A, B or C, is the reaction fastest?

b What is it about the line on the graph which helped you answer part **a**?

c At which point, A, B or C, has the reaction reached the endpoint?

d What is it about the line on the graph which helped you answer part **c**?

e From the graph, work out the total loss in mass during the reaction.

f Predict the mass of the flask and contents after 6 minutes.

3 Chemical structure

This chapter includes coverage of:

N3 Chemical structure

CL3 Materials SCN 3-16b

Exercise 3A Substances and their states

1. Complete the following sentence.

   The three states of matter are s _ _ _ _ , l _ _ _ _ _ and g _ _.

2. Use the word bank to help you complete the following sentences.

   (The words may be used more than once.)

saturated	soluble	solute	solution	solvent

   a A substance which dissolves in water is said to be _ _ _ _ _ _ _.

   b A _ _ _ _ _ _ is a substance which dissolves.

   c The liquid in which a substance dissolves is called the _ _ _ _ _ _ _.

   d A _ _ _ _ _ _ _ _ is formed when a substance dissolves in a liquid.

   e When no more of a substance will dissolve in a liquid the _ _ _ _ _ _ _ _ is said to be
   _ _ _ _ _ _ _ _.

3. When carbon dioxide is bubbled through water, carbonic acid is formed.

   Name:

   a the solvent

   b the solution

   c the solute.

4. Some students dissolved different substances in water. Here is the report one student wrote.

   We measured out 50 cm³ of water into three different boiling tubes. We added different chemicals to each boiling tube and stirred each one until the chemicals dissolved. We did this until no more chemical would dissolve. We were able to dissolve three spoonfuls of copper sulfate, four spoonfuls of sodium chloride and nine spoonfuls of sugar.

   a Suggest what the aim of the experiment was.

   b Present the results the students obtained in a table. Use the headings 'Chemical' and 'Number of spoonfuls dissolved'.

   c State why it was important to use the same volume of water in each experiment.

   d Suggest what the students would have seen which would indicate they should stop adding more spoonfuls of chemical.

   e What name is given to a solution which can't dissolve any more chemical?

   f State **one** conclusion the students could draw from the results of their experiments.

Exercise 3B Elements, compounds and mixtures

This exercise includes coverage of:

N3 Chemical structure

CL3 Materials SCN 3-15b, 3-16a, 3-16b

1 Use the word bank to help you complete the following statements.

compounds joined elements atoms mixture

a Everything is made up of _ _ _ _ _.

b _ _ _ _ _ _ _ _ are made up from one kind of atom.

c _ _ _ _ _ _ _ _ _ are made up of different atoms _ _ _ _ _ _ together.

d In a _ _ _ _ _ _ _ of elements the atoms are not joined.

2 Here is a student report of an experiment they carried out.

We mixed some iron and sulfur together in a dish. We then held a magnet over the mixture and the iron stuck to the magnet. We then heated the iron and sulfur mixture with a Bunsen burner. After the mixture had been heated for a while the iron didn't separate out when the magnet was held over it.

Explain why the iron could not be separated out using the magnet after the mixture was heated.

3 Filtering and evaporation are two ways of separating substances. Choose the best way to separate the following:

a sand and water

b water from the salts dissolved in seawater

c alcohol and water in a distillery

d drinking water from leaves and other solid particles in a reservoir.

4 Diagrams **A–D** show atom models.

A   B   C   D

a Which model represents an element?

b Which model represents a mixture of compounds?

c Which two models represent pure compounds?

5 Copy the table and complete it.

Elements reacting	Name of compound formed
sodium and iodine	
silicon and carbon	
magnesium and chlorine	
potassium and sulfur	
	lead bromide
	magnesium oxide

6 Write word equations for the reactions taking place in Question 5.

Exercise 3C Elements and the periodic table

This exercise includes coverage of:

N3 Chemical structure

CL3 Materials SCN 3-15a

1 Complete the following sentences using the words in the word bank to help you.

atomic	atoms	group	periodic	symbol

   a Elements are made from the same _ _ _ _ _.

   b The elements are arranged in the _ _ _ _ _ _ _ _ table.

   c Elements with similar chemical properties are in the same vertical _ _ _ _ _.

   d Each element has its own _ _ _ _ _ _ and _ _ _ _ _ _ number.

2 The diagram shows part of the periodic table.

   a Give the symbol for the element which has similar chemical properties to sodium (Na).

   b Give the symbol for the element which is least reactive.

   c Give the symbol for the non-metal which exists in a form which conducts electricity.

   d State which group is also known as the alkali metals.

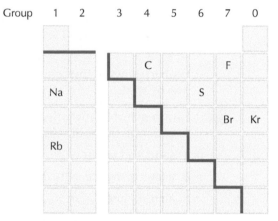

3 Use a full periodic table to identify:

   a i the element with atomic number 29

     ii Is the element a metal or a non-metal?

   b i the name and atomic number of the element with the symbol I

     ii Is the element a metal or non-metal?

   c Describe a test you could carry out to prove if the elements in parts **a** and **b** were metal or non-metal.

    Include a diagram of the arrangement you would use.

4 The table shows the properties of some elements (the letters are not the symbols of the elements).

Element	Properties
X	very reactive non-metal
Y	very unreactive non-metal
Z	very reactive metal

   a Which element is:

     i a noble gas   **ii** a halogen?

   b i Suggest how element Z would be stored in the laboratory.

     ii To which group in the periodic table does element Z belong?

Exercise 3D Chemical formulae of compounds

This exercise includes coverage of:

N3 Chemical structure

CL3 Materials SCN 3-15b

Hint | **Bonding arms**

The idea of bonding arms can be used to work out the chemical formulae for compounds.

Group	1	2	3	4	5	6	7	0
Number of bonding arms	1	2	3	4	3	2	1	0

Example 3.1: magnesium bromide

Bonding arms:

magnesium (Mg): group 2, so 2 bonding arms

bromine (Br): group 1, so 1 bonding arm

The two bonding arms of the magnesium need two arms to join with, so the magnesium joins with two bromine atoms.

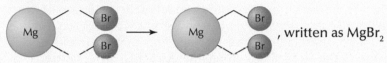

, written as $MgBr_2$

 1 Use bonding arms to work out the formulae of the following compounds. Element symbols can be found in the periodic table.

    a Hydrogen fluoride

    b Calcium oxide

    c Potassium sulfide

    d Magnesium chloride

    e Sodium nitride

Hint | **Prefixes**

Prefixes in the name of a compound can tell you how many atoms of an element are in the formula.

Some prefixes and their meaning are shown in the table.

Prefix	mono-	di-	tri-	tetra-	penta-
Number	1	2	3	4	5

Example 3.2: carbon tetrachloride

tetra = 4, so one carbon is joined to four chlorines; the formula is CCl_4

 2 Work out the formulae for the following compounds. Element symbols can be found in the periodic table.

    a Iodine monochloride   **b** Nitrogen dioxide

    c Nitrogen trifluoride   **d** Silicon tetrachloride

    e Phosphorus pentabromide

4 Atomic structure and bonding related to properties of materials

Exercise 4A Atomic structure

1 **a** Copy and complete the table, which shows the particles which make up an atom and their charge and mass.

b **i** Draw a diagram of the atom and label it to show where the particles in the table in part **a** are found.

ii On the diagram label the nucleus of the atom.

Particle	Charge	Mass
proton (_)	—	–
____ (e⁻)	—	almost zero
____ (_)	no charge	–

2 Read the following paragraph and answer the questions below.

In the early 20th century JJ Thomson's 'plum pudding' model of the structure of the atom proposed that the positive and negative particles in atoms were spread out through the atom. Ernest Rutherford later put forward the idea that the positive particles were at the centre of the atom and the negative particles moved around the positive particles. In 1932 James Chadwick proved that there was a third particle in an atom which had the same mass as a positive particle but had no charge.

a Name the positive and negative particles referred to in the paragraph.

b **i** Name the particle discovered by Chadwick.

ii State where in the atom this particle is found.

3 **a** State what is meant by the **atomic number** of an element.

b Explain how atoms can be neutral if they contain positive and negative particles.

c State what the **mass number** of an element is.

4 The table below gives information about the atoms of three elements. Use a periodic table to help you complete entries **a–o**.

Element	Symbol	Atomic number	Protons	Electrons	Neutrons	Mass number
fluorine	a	b	c	d	10	e
f	g	h	19	i	j	39
k	l	36	m	n	44	o

5 The table gives the number of protons and neutrons in atoms of different elements.

Element	Number of protons	Number of neutrons
lithium	3	4
phosphorous	15	16
zinc	30	35
zirconium	40	51

From the information in the table, write a statement comparing the number of protons and neutrons in an atom.

Exercise 4B Bonding

1 Non-metal elements and compounds made from non-metal atoms exist as groups of atoms bonded together.

  a What name is given to these groups of atoms bonded together?

  b What name is given to the bond formed between non-metal atoms?

  c How is the bond between non-metals formed?

  d Use 'dot and cross' diagrams to show how bonds are formed between:

    i two fluorine atoms

    ii a hydrogen atom and a chlorine atom.

> **Hint** Each fluorine atom has seven electrons in its outer shell. A hydrogen atom has one electron in its outer energy level and a chlorine atom has seven.

  e Use a line between the symbol for each element to represent the bonds in a fluorine molecule and a hydrogen chloride molecule.

2 When metal atoms react with non-metal atoms, charged atoms are formed.

  a What name is given to these charged atoms?

  b What type of charge is formed on a metal atom?

  c What type of charge is formed on a non-metal atom?

  d Explain how these charged atoms are formed.

3 When sodium reacts with chlorine, sodium chloride is formed.

  a Name the type of bond formed when sodium and chlorine atoms react.

  b Describe how the bond in part **a** is formed.

4 Complete entries **a–h** in the table.

Element	Symbol	Number of electrons in atom	Electrons lost or gained	Ion symbol	Number of electrons in ion
a	Li	**b**	loses 1e⁻	**c**	**d**
sulfur	**e**	**f**	**g**	S^{2-}	**h**

5 What type of bonding is found in the following compounds?

  a Lithium iodide (LiI)

  b Magnesium oxide (MgO)

  c Nitrogen trichloride (NCl_3)

  d Sulfur dioxide (SO_2)

Exercise 4C Properties of covalent and ionic substances

1 The table shows some properties of four substances, A–D.

Substance	Melting point (°C)	Boiling point (°C)	State at room temperature (20 °C)
A	−78	−33	
B	−7	59	
C	2614	2850	
D	961	1560	

a Use the melting point and boiling point of each substance to work out the state at which it exists at room temperature.

b i Which substances are likely to exist as covalent molecules?

ii Which of the substances are ionic?

c Copy and complete the general statements about covalent molecular and ionic substances.

i Covalent substances generally have _ _ _ melting points and boiling points and exist mainly as _ _ _ _ _ and _ _ _ _ _ _ _.

ii Ionic substances generally have _ _ _ _ melting points and boiling points and exist as _ _ _ _ _ _.

d Substance B is a halogen. Use a periodic table or a data booklet to help you identify the element.

2 Copy and complete the table below using the word bank to help you.

atoms	attract	compounds	electricity	gain	gases	ions
lose	magnesium oxide	molecules	negatively	non-metal	pairs
positively	solids	solution	water

Ionic bonding	Covalent bonding
Found in _ _ _ _ _ _ _ _ _ formed between metals and non-metals. Examples: Sodium chloride and _ _ _ _ _ _ _ _ _ _ _ _ _ _ _	Found in elements and compounds made up of _ _ _ _ _ _ _ _ _ atoms. Examples: chlorine and _ _ _ _ _
Formed when metal atoms _ _ _ _ electrons and form _ _ _ _ _ _ _ _ _ _ charged _ _ _ _. Non-metal atoms _ _ _ _ electrons to form _ _ _ _ _ _ _ _ _ _ charged ions. Oppositely charged ions _ _ _ _ _ _ _ each other and form a giant crystal structure.	Formed when _ _ _ _ _ of electrons are shared between non-metal _ _ _ _ _. Exist as individual _ _ _ _ _ _ _ _ _.
All exist as _ _ _ _ _ _ at room temperature. Conduct _ _ _ _ _ _ _ _ _ _ _ when in _ _ _ _ _ _ _ _ or melted.	Can exist as _ _ _ _ _, liquids and solids at room temperature.

3 Groups of students were investigating the bonding in some solid compounds. They decided to test the electrical conductivity of each substance.

a Describe, with the aid of a diagram, how they could test the electrical conductivity of the substances.

b State how they would know if a substance was a conductor or not.

c One group found that none of the solids conducted so concluded they must all be covalent substances.

Comment on their conclusion and state what they should have done to make sure their conclusion was accurate.

4 Water has a melting point of 0 °C and a boiling point of 100 °C.

a i Suggest the type of bonding which exists in water.

ii Explain your answer to part **i**.

b Water can exist as a solid, liquid or gas depending on the temperature.

In which state will water exist at:

i 120 °C

ii –7 °C

iii 20 °C?

c Give the state symbols for solid, liquid and gas.

d When sugar dissolves in water a solution is formed.

Write the state symbol for a solution formed when water is the solvent.

Exercise 4D Formulae of elements and compounds

1 Give the formula of each of the following elements.

a Potassium

b Calcium

c Bromine

d Fluorine

e Argon

f Nitrogen

2 Copy and complete the following table.

Name of compound	Elements present in compound
calcium chloride	
iron sulfate	
zinc carbonate	
copper nitrite	
potassium sulfite	
magnesium nitrate	

3 Work out the chemical formulae of the following compounds from their molecular models and structural formulae.

a

b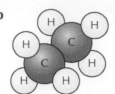

c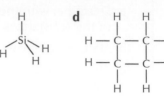

d

4 Use valency rules to work out the formula of each compound.

   a Magnesium chloride   **b** Hydrogen fluoride

   c Aluminium hydride   **d** Barium sulfide

   e Magnesium carbide

5 Write chemical symbol equations for the following word equations.

   a magnesium + chlorine → magnesium chloride

       (element)   (diatomic   (compound)

                  element)

   b hydrogen   + iodine   → hydrogen iodide

   c lithium   + sulfur   → lithium sulfide

   d carbon   + oxygen → carbon dioxide

> **Hint** Identify the elements and compounds in a word equation.
>
> Elements have their symbol as their formula, except for the diatomic elements, e.g. Cl_2.
>
> The formula of a compound can be worked out using valency.

6 Using relative atomic mass (RAM) values from a data book, calculate the formula mass of the following.

   a Chlorine (Cl_2)

   b Carbon monoxide (CO)

   c Potassium bromide (KBr)

   d Sodium sulfide (Na_2S)

   e Aluminium oxide (Al_2O_3)

7 The table shows the relative atomic masses and melting points of some of the group 1 elements.

Element symbol	Relative atomic mass (RAM)	Melting point (°C)
Li	7	180
Na	23	98
K	39	X
Rb	85·5	39

   a From the information in the table predict the melting point (X) of potassium.

   b Write a general statement linking relative atomic mass and melting point for the group 1 metals.

5 Energy changes of chemical reactions

Exercise 5A Exothermic and endothermic reactions

1 A group of students carried out some experiments to investigate energy changes. Here is one of their reports.

In the first experiment we mixed some acid and alkali. The temperature of the mixture went up. In the second experiment potassium chloride was dissolved in water. The temperature of the solution formed was lower than the temperature of the water.

   a In which experiment was energy given out?

   b Explain your answer to part **a**.

   c What term is used to describe a chemical reaction in which energy is given out?

   d What term is used to describe a chemical reaction in which energy is taken in from the surroundings?

2 Three reactions involving solutions were carried out. The results are shown in the table.

Reaction	Temperature of reactants before mixing (°C)	Temperature after mixing (°C)
1	20	23
2	18	22
3	22	20

   a i State whether each reaction is exothermic or endothermic.

     ii Explain each of your answers in part **i**.

   b i Which reaction is the most exothermic?

     ii Explain your answer to part **i**.

3 The diagram shows a self-warming drinks can.

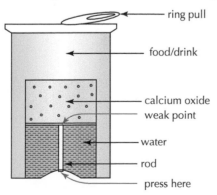

   a Suggest what happens inside the can to produce the heat which warms the drink.

   b What name is given to a reaction which produces heat?

4 A student added ammonium nitrate to a conical flask containing water. Some of the water had spilled onto the work bench and got under the flask. When the student tried to move the flask he found it was frozen to the bench.

   a Suggest what happened to cause the flask to freeze to the bench.

   b What type of energy change had taken place?

5 A group of students wanted to investigate whether the reaction between hydrochloric acid and sodium hydroxide was exothermic or endothermic.

   a State the aim of the experiment.

   b Write out a plan they could follow. Include the method they should use and any measurements they should make.

   c State how the students would know if the reaction was exothermic or endothermic.

6 **a** State whether the following processes are exothermic or endothermic.

   i Plants using the Sun's energy during photosynthesis

   ii Burning gas in a cooker

   iii Mixing chemicals in a cold pack used to treat sports injuries

   iv Gunpowder exploding in a firework

   b Explain each of the answers you gave in part **a**.

7 Thermal runaway is a process which can lead to explosions in industrial sites. It is caused by a build-up of heat during a chemical reaction.

   What name is given to chemical reactions which give out heat?

6 Acids and bases 1

This chapter includes coverage of:

N3 Acids and bases

CL3 Materials SCN 3-18a

Exercise 6A Acids and alkalis

1 Copy and complete the following sentences. Use the word bank to help you.

| above | below | exactly | zero | 14 |

a The pH scale is a numbered scale from below _ _ _ _ to above _ _.

b Acids have a pH _ _ _ _ _ 7.

c Alkalis have a pH _ _ _ _ _ 7.

d Neutral solutions have a pH of _ _ _ _ _ _ _ 7.

2 The chart shows the pH range covered by universal indicator.

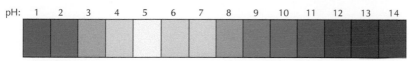

pH: 1 2 3 4 5 6 7 8 9 10 11 12 13 14

a State what colour the indicator would be if a solution is pH 1.

b State what colour the indicator would be if a solution is pH 14.

c Copy and complete the following sentences by choosing the correct word in brackets.

  i A solution with pH 1 is (**more** / **less**) acidic than a solution with pH 6.

  ii A solution with pH 8 is (**more** / **less**) alkaline than a solution with pH 14.

3 The table shows some of the properties of water, bicarbonate of soda and vinegar. Copy and complete the table.

Substance	Acid/Alkali/Neutral	pH below 7 / 7 / above 7	Colour of universal indicator
			red
		7	
	alkali		

4 The table shows the pH of some everyday solutions.

a i Which two solutions are acids?

  ii Which of the solutions is more acidic?

b i Which two solutions are alkalis?

  ii Which of the solutions is more alkaline?

c i Suggest what colour universal indicator would be when lemon juice is added.

  ii Suggest what colour universal indicator would be when oven cleaner is added.

Solution	pH
lemon juice	2
milk of magnesia	11
oven cleaner	14
tea	5

5 Some plants can be used as indicators. The table shows the colour changes when some plant indicators have acid or alkali added.

Plant used to make indicator	Colour of indicator	Colour in acid	Colour in alkali
red cabbage	purple	red	blue
beetroot	red	red	purple
cherries	red	red	blue

A number of solutions were tested with the indicators listed in the table.

a The indicator made from cherries turned blue when solution 1 was added.

What does this indicate about the solution?

b The indicator made from beetroot stayed red when solution 2 was added.

Explain why this doesn't necessarily mean that solution 2 is acidic.

c i Suggest which of the indicators in the table is the most useful.

ii Explain your answer to part **i**.

6 The table shows how the pH of milk changes when it is left out of the fridge.

Number of days	pH
1	6·5
2	6·0
3	5·5
4	

a Copy and complete the following sentence by choosing the correct word in brackets.

As the days go by the milk becomes more (**acidic** / **alkaline**).

b Predict the pH of the milk after 4 days.

7 Copy and complete the following sentences by choosing the correct word in brackets.

a When water is added to acid, the acid becomes more (**dilute** / **concentrated**) and the pH of the acid (**increases** / **decreases**) towards 7.

b When water is added to alkali, the alkali becomes more (**dilute** / **concentrated**) and the pH of the alkali (**increases** / **decreases**) towards 7.

8 The 'fizz' in drinks is caused by an acidic gas called carbon dioxide.

a Suggest what the pH of a fizzy cola drink might be.

b Suggest why a dentist would be concerned about people drinking too many fizzy drinks.

9 Acids are found naturally in foods. They are also often added by food manufacturers.

a Name a food which contains naturally occurring citric acid.

b Bread baked with added propionic acid takes longer to go mouldy than the same bread baked without propionic acid.

Suggest what the propionic acid is acting as.

Exercise 6B Neutralisation

This exercise includes coverage of:

N3 Acids and bases

CL3 Materials SCN 3-18a

1. Some acid was spilled in the laboratory. Before cleaning it up, the teacher first poured water on the acid and then some alkali.

   a Explain why the teacher added water to the acid.

   b What happens to the pH of an acid when it reacts with an alkali?

   c What happens to the pH of an alkali when it reacts with an acid?

   d Name the type of reaction which takes place when alkali is added to acid.

   e Name the two products which are formed when an acid and alkali react.

2. The reaction of an acid with an alkali can be followed by slowly adding the acid to a conical flask containing alkali and universal indicator and noting the colour changes.

   a Suggest what colour the indicator would be in the alkali.

   b Suggest a pH number for the alkali.

   c What happens to the colour of the indicator as acid is added?

   d What happens to the pH of the alkali as the acid is added?

   e What colour will the indicator be when the pH reaches 7?

   f What name is given to a solution with pH 7?

3. The table shows some common acids and the name endings of the compounds formed when they react with alkalis.

Acid	Salt name ending
hydrochloric	chloride
sulfuric	sulfate
nitric	nitrate

   a Water is produced when each of the acids in the table react with magnesium hydroxide. Name the other compound produced in each reaction.

   b What general name is given to the compounds formed in part a?

4. Copy and complete each of the following word equations.

   a potassium hydroxide + hydrochloric acid → water + _____ _____

   b lithium hydroxide + sulfuric acid → water + _____ _____

   c calcium hydroxide + _____ _____ → _____ + calcium nitrate

5. Vinegar is an acid which can be applied to the skin to treat a wasp sting.

   a What does this suggest about the chemical in a wasp's sting?

   b Name the type of reaction taking place between the vinegar and the chemical in the wasp's sting.

6. Indigestion is caused by too much acid in the stomach. Indigestion tablets contain chemicals which reduce the acidity.

   a State what kind of chemicals are likely to be in an indigestion tablet.

   b Name the type of reaction taking place between the acid and the chemical in an indigestion tablet.

Exercise 6C Acidic gases and the environment

> **This exercise includes coverage of:**
>
> **N3** Acids and bases
>
> **CL3** Planet Earth SCN 3-05b
>
> **CL3** Materials SCN 3-18a

1 Rainwater is naturally acidic due to naturally occurring gases. Other gases produced by human activity are making rainwater even more acidic.

   a Name the naturally occurring gas which contributes most to rainwater being acidic.

   b Name **two** of the other gases produced by human activity which contribute to the formation of acid rain.

   c State the main source of the gases which cause acid rain.

   d Describe the effect acid rain has on:

     i trees

     ii fish in lochs.

   e Suggest **one** way acidification of lochs has been treated.

   f Suggest **one** way of reducing acid rain.

2 Some students wanted to test the pH of rainwater.

   a Describe how they could find the pH of rainwater.

   b One student suggested testing rainwater from the roof of the school which had been collected in a water butt.

     i Suggest why this might not give an accurate result.

     ii Suggest a method of collecting a pure sample of rainwater to test.

3 Between 1997 and 2017, the amount of sulfur dioxide produced in the European Union (EU) decreased. 20% of this decrease was from energy used in industry, 15% from homes and businesses, 5% from road transport and 5% from manufacturing. The biggest decrease was from energy production.

   a Calculate the decrease in sulfur dioxide production in the energy production sector.

   b Present the information on the decrease in sulfur dioxide emissions in a table. Use the headings 'Source of sulfur dioxide' and 'Percentage reduction'.

   c The pie chart shows the reduction in sulfur dioxide emissions from the sources mentioned above.

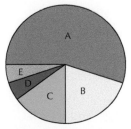

   Match each label in the pie chart to the correct source of reduction mentioned in the passage.

4 The graph shows how the amount of nitrogen dioxide produced in the UK has been changing since 1990.

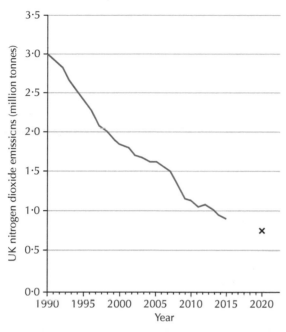

a Describe the trend in nitrogen dioxide emissions from 1990 to 2015.

b Point X on the graph shows the UK's target for nitrogen dioxide emissions.

   i By which date is the target to be reached?

   ii Predict whether the UK is on course to meet its target.

5 The table shows the amount of nitrogen oxides being released into the atmosphere by a country in the period 1970–2010.

Year	Mass of nitrogen oxides produced (million tonnes)
1970	6·5
1980	5·0
1990	3·5
2000	1·5
2010	0·5

a Draw a bar chart of 'Year' against 'Mass of nitrogen oxides produced'.

b Describe the trend in nitrogen oxide emissions.

c Predict the number of tonnes of nitrogen oxides produced in 2020 if the trend continues.

d Road transport is responsible for much of the nitrogen oxides produced.

   Suggest what would happen to the nitrogen oxides emissions from vehicles if everyone drove electric vehicles.

6 Our seas and oceans dissolve about 25% of the carbon dioxide produced by human activity.

a State **one** of the benefits of our seas and oceans dissolving carbon dioxide.

b State **one** disadvantage of our seas and oceans dissolving carbon dioxide.

Exercise 6D Greenhouse gases and global warming

This exercise includes coverage of:

N3 Acids and bases

CL3 Planet Earth SCN 3-05b

1 Copy and complete the summary, using the word bank to help you.

| atmosphere | climate | greenhouse | heat | rise | warming |

Carbon dioxide is one of the gases in the _ _ _ _ _ _ _ _ _ which is thought to be preventing _ _ _ _ from the Earth escaping into space. This is known as the _ _ _ _ _ _ _ _ _ _ effect. This effect is causing the temperature of the Earth to _ _ _ _. This is known as global _ _ _ _ _ _ _ which many scientists think is contributing to _ _ _ _ _ _ _ change.

2 **a** State **two** possible effects of climate change.

   b Give **one** example of how we can prevent more carbon dioxide getting into the atmosphere.

3 The graph shows how the percentage of carbon dioxide in the atmosphere has changed since 1860.

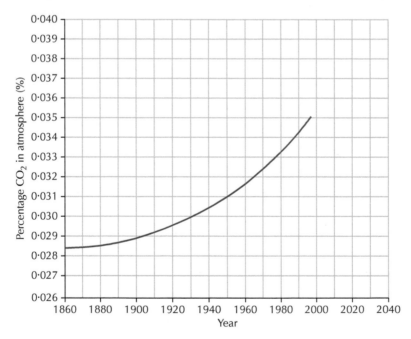

a State the general trend in the change in atmospheric carbon dioxide levels since 1860.

b In which year did the level of carbon dioxide in the atmosphere reach 0·030%?

c **i** From the graph predict what the percentage of carbon dioxide in the atmosphere will be in 2040 if it continues to change at the same rate.

   ii Using your answer to part **i** suggest what effect this will have on the temperature of the Earth.

7 Acids and bases 2

This chapter includes coverage of:

N4 Acids and bases

Exercise 7A Bases

1 Copy and complete the following summary. Use the word bank to help you.

above	alkaline	base	dissolve	neutralisation	salt	water

A _ _ _ _ is a compound which can react with an acid to form _ _ _ _ _ and a
_ _ _ _.

This is called _ _ _ _ _ _ _ _ _ _ _ _ _ _ _. Soluble bases can _ _ _ _ _ _ _ _ in water to
form _ _ _ _ _ _ _ _ solutions. The pH of water changes from 7 to _ _ _ _ _ 7.

2 **a** Select the bases from the following list.

    i Potassium oxide   **ii** Magnesium carbonate   **iii** Calcium chloride

    iv Sodium hydroxide   **v** Lithium iodide   **vi** Barium fluoride

  b Use a solubility table in a data booklet to say which bases you identified in part **a**
would form alkaline solutions.

3 The table gives the names and formulae of some acids.

  a Suggest which element gives each acid its acidic
properties.

  b Name the salt formed when the following bases react
with each of the acids in the table.

Acid	Formula
hydrochloric	HCl
nitric	HNO_3
sulfuric	H_2SO_4

    i Copper oxide   **ii** Barium hydroxide   **iii** Magnesium carbonate

4 **a** Copy and complete the following word equations.

    i nickel oxide   + nitric acid   → water   + _____ _____

    ii magnesium hydroxide + sulfuric acid   → _____   + _____ _____

    iii barium carbonate   + hydrochloric acid → _____ + _____ _____
                                                            + _____ _____

    iv _____ _____   + _____ acid   → _____ + zinc nitrate
                                                                   + carbon dioxide

  b Name the type of reaction taking place in all of the reactions in part **a**.

5 The table shows what happens to the pH of water
when two oxides, X and Y, are added to water.

  a Copy and complete the following sentence
by choosing the correct word in the brackets.

    Oxides X and Y are both (**insoluble / soluble**).

Oxide	pH of water	pH of water after oxide added
X	7	9
Y	7	3

  b **i** Which oxide is a metal oxide?

    ii Explain your answer to part **i**.

  c Nickel oxide has no effect on the pH of water.

    What does this tell us about the solubility of nickel oxide?

6 Dilute hydrochloric acid was added to calcium carbonate.

a Name the salt formed in the reaction.

b Name the type of reaction taking place.

c What evidence would indicate that all the acid had reacted?

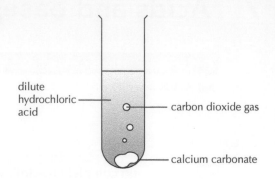

dilute hydrochloric acid

carbon dioxide gas

calcium carbonate

Exercise 7B Gases and the environment

This exercise includes coverage of:

N4 Acids and bases

CL4 Materials SCN 4-18a

Hint   Make sure you can answer the questions in Exercises 6C and 6D before answering these questions.

1 The manufacture of cement is considered to be a major source of a gas which has a negative effect on the environment. The main raw material used to make cement is calcium carbonate. Fossil fuels have traditionally been used to produce the energy needed in the process.

a i Name the gas produced during the manufacture of cement.

ii Suggest another source of the gas you gave as your answer in part i.

iii Suggest what could be done in cement works to stop polluting gases reaching the atmosphere.

b Fuels such as old tyres can be used instead of fossil fuels to produce the energy needed to make cement. They reduce the carbon footprint of the process.

Explain what is meant by **carbon footprint**.

2 The graph shows sulfur dioxide emissions in the UK since 1990.

a Describe the general trend in UK sulfur dioxide emissions.

b The **o** in the graph indicates the UK target for sulfur dioxide emissions in 2010.

Does the graph indicate whether the UK reached its target or not?

c The **x** on the graph indicates the UK target for sulfur dioxide emissions in 2020.

Does the graph indicate whether the UK will reach its target or not?

d State the main source of sulfur dioxide emissions in the UK.

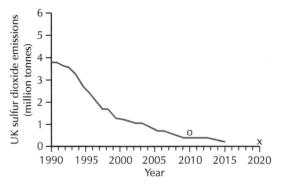

8 Fuels and energy 1: Fuels

> **This chapter includes coverage of:**
>
> **N3** Fuels and energy
>
> **CL3** Planet Earth SCN 3-05b

Exercise 8A Energy from fuels

1 Energy is used in industry to make useful materials such as steel.

Give **three** everyday ways in which we use energy.

2 We get much of our energy by burning substances.

  a Give the name used to describe any substance we can burn to give us energy.

  b Copy and complete the following sentence by choosing the correct word in the brackets.

    The main type of energy given out when a substance burns is (**heat / light**).

  c Substances which burn react with a gas in the air. Name the gas.

3 Fossil fuels are our main source of energy.

  a i State what is meant by a **fossil fuel**.

    ii Copy and complete the following sentence by choosing the correct word from the brackets.

      Fossil fuels are (**natural / synthetic**).

  b Crude oil is an example of a fossil fuel.

    i Name **three** other fossil fuels.

    ii Name **two** fuels obtained from crude oil which are used as fuels in road vehicles.

    iii Crude oil has to be heated to separate out the mixture of liquid fuels it contains.

      Name the process used to separate a mixture of liquids.

4 Fossil fuels are mainly made up of compounds called hydrocarbons.

  a State what is meant by a **hydrocarbon**.

  b The diagrams show models of some hydrocarbons.

  Work out the chemical formula of each compound.

  i

hydrogen
carbon

  ii
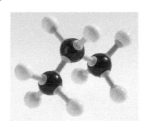

(continued)

iii 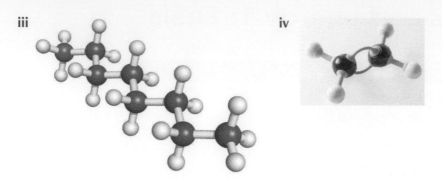 **iv**

5 **a** State which of the following chemical formulae represent a hydrocarbon.

    i C_2H_6 **ii** C_4H_8 **iii** C_2H_5OH **iv** C_6H_{14} **v** CH_3Cl

    b State why the formulae you didn't choose in part **a** do not represent a hydrocarbon.

6 The apparatus shown can be used to identify the products when hydrocarbons burn.

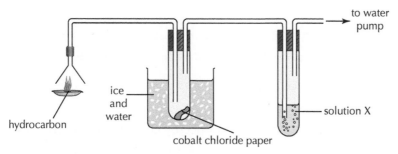

    a Carbon dioxide gas is one of the products.

      i State the name of solution X which is used to prove that carbon dioxide is produced.

      ii Describe what you would see happening to solution X when the carbon dioxide bubbles through it.

    b The other product is water.

      i State what would happen to the cobalt chloride paper to indicate that water is produced.

      ii State one other test which would confirm water was produced.

      iii Suggest why the test tube which collected the water is in an ice-water bath.

    c **i** Write a word equation for a hydrocarbon reacting with oxygen from the air to produce carbon dioxide and water.

      ii Identify each substance in the word equation as either a compound or element.

      iii Use the bonding arms method to work out the chemical formula for water (hydrogen oxide).

      iv Write the chemical formula for carbon dioxide.

7 A student gently blew through a straw into limewater. The limewater turned cloudy.

    a State which gas this shows is present in the air we breathe out.

    b Water and the gas produced in part **a** are waste products which come from the food we eat when it reacts with oxygen in the cells of the body.

    State what else is produced during this process.

8 A group of students were investigating how much energy we get from foods. They did this by comparing the temperature rise of water when each food is burned. The diagram shows the apparatus they used.

They used a sweet biscuit (sugar), peanuts (vegetable oil), bread (starch) and lean meat (protein).

They measured the rise in the temperature of the water when the same weight of each food was burned.

The results are shown in the table.

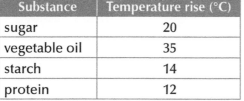

Substance	Temperature rise (°C)
sugar	20
vegetable oil	35
starch	14
protein	12

a State the aim of the experiment.

b i Which substance produced most energy?

   ii Explain your answer to part **i**.

c State why it is important to burn the same mass (weight) of food.

d Name **one** other variable which should be kept the same in the experiment.

e It was suggested that the group should carry out each experiment three times.

  Explain why this should be done.

f Copy and complete the following general conclusion by choosing the correct word in the brackets.

  (**All** / **One** / **Some**) of the substances tested produced energy when burned.

9 Alcohols can be burned to give us energy.

a A model of the alcohol called ethanol is shown.

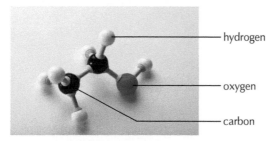

Write the chemical formula for ethanol.

b A student said that carbon dioxide and water are formed when ethanol burns so it must be a hydrocarbon.

  i State what is wrong with the student's conclusion.

  ii Describe a chemical test for carbon dioxide.

  iii Describe **two** tests which could be carried out to prove that water is formed.

c Write the word equation for ethanol burning to form carbon dioxide and water.

9 Fuels and energy 2: Controlling fires

This chapter includes coverage of:

N3 Fuels and energy

Exercise 9A The fire triangle

1 In order for burning to take place oxygen is needed.

    a State where the oxygen comes from.

    b Name the other two conditions needed for burning to occur.

    c Draw a fire triangle showing the three conditions needed in order for burning to take place.

2 The organisers of a bonfire have to decide how they are going to put the fire out at the end of the night.

    a Copy out each suggestion and complete it by choosing the correct word in the brackets.

      i Pour water on the fire to remove the (**fuel / heat / oxygen**).

      ii Shovel earth onto the fire to remove the (**fuel / heat / oxygen**).

    b State why, if just left, the fire would eventually go out.

3 The table gives information about different types of fire extinguisher.

Extinguisher		Type of fire					
Colour	Type	Solids (wood, paper, cloth, etc)	Flammable liquids	Flammable gases	Electrical equipment	Cooking oils & fats	Special notes
	Water	✓ Yes	✗ No	✗ No	✗ No	✗ No	Dangerous if used on 'liquid fires' or live electricity.
	Foam	✓ Yes	✓ Yes	✗ No	✗ No	✓ Yes	Not practical for home use.
	Dry powder	✓ Yes	✓ Yes	✓ Yes	✓ Yes	✗ No	Safe use up to 1000 V.
	Carbon dioxide (CO_2)	✗ No	✓ Yes	✗ No	✓ Yes	✓ Yes	Safe on high and low voltages.

    a Use the information in the table to select the fire extinguisher which would be best for putting out:

      i a chip pan in the kitchen at home   **ii** an electrical fire

      iii the gas fire in a gas barbeque   **iv** an armchair on fire in the living room.

    b Suggest why water should never be used to put out an electrical fire.

    c Suggest why different types of fire extinguishers have labels with different coloured backgrounds.

4 Copy and complete the following sentences by choosing the correct word in the brackets.

a Using a fire blanket to cover a chip pan which is on fire removes the (**fuel** / **heat** / **oxygen**).

b Spraying a layer of powder over a fire removes the (**fuel** / **heat** / **oxygen**).

c A water extinguisher removes the (**fuel** / **heat** / **oxygen**).

d Carbon dioxide from an extinguisher removes the (**fuel** / **heat** / **oxygen**).

5 Experts say that a fire in a tower block which killed many people might have been prevented from spreading if a water sprinkler system had been installed.

State how water stops a fire from burning.

6 Firefighters in California used a mixture of water and a chemical which made the water sticky to try to stop a wildfire spreading. High winds at the time made the fire burn even more fiercely.

a State what water does to stop fires burning.

b Suggest why the water being sticky makes it better at putting out fires than water on its own.

c Copy and complete the following sentences by choosing the correct word in the brackets.

   i High winds supply extra (**fuel** / **oxygen** / **heat**).

   ii The chemical in the sticky water also helps stop (**heat** / **oxygen**) reaching the fuel.

   iii If left to burn, a wildfire will not stop until it runs out of (**heat** / **oxygen** / **fuel**).

10 Fuels 1: Fossil fuels

Exercise 10A Formation of fossil fuels and new technologies

 1 The diagram shows how plants can trap the Sun's energy.

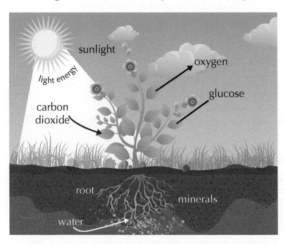

a Copy and complete the word equation using the information in the diagram to help you.

_____ _____ + _____ → glucose + _____

b State where the energy needed for the reaction comes from.

c What name is given to a chemical reaction which takes in energy?

d State in which part of a compound the energy taken in by a plant is stored.

e Name the green chemical in plants which is needed to trap energy.

f Name the overall process taking place.

2 Copy and complete the following paragraphs, which are about how plants provide us with energy. Use the word bank to help you.

animals	bonds	coal	decayed	Earth	energy	fossil	fuel	gas
heat	millions	oil	photosynthesis	plants	Sun's	weight	wood	

Trees are _ _ _ _ _ _ which provide us with _ _ _ _ which can be burned to give us _ _ _ _ _ _. Plants trap the _ _ _ _ energy in chemical _ _ _ _ _ through a process called _ _ _ _ _ _ _ _ _ _ _ _ _ _.

Even when plants die they can still be used as a _ _ _ _. For example, _ _ _ _ was formed from trees and other plants which lived and died _ _ _ _ _ _ _ _ of years ago. When they died their remains partly _ _ _ _ _ _ _ (rotted) and were gradually covered in layers of mud and sand. The _ _ _ _ _ _ of these layers combined with the _ _ _ _ from within the _ _ _ _ _ and gradually changed the plant material into coal.

Crude _ _ _ and natural _ _ _ were formed in a similar way but from dead sea _ _ _ _ _ _ _ and plants.

These fuels are known as _ _ _ _ _ _ fuels.

3 **a** Fossil fuels are described as finite resources.

Explain what is meant by **finite resource**.

b The UK has reserves of oil under the North Sea, discovered in the early 1970s. The graph shows UK oil production from 1975 until 2017.

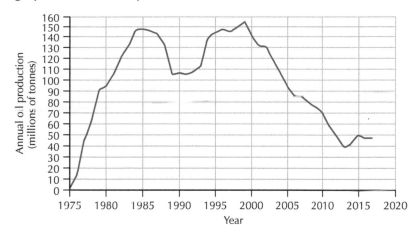

Source: Department of Energy and Climate Change

i Suggest why only about 1 million tonnes of oil was produced from UK oil fields in 1975.

ii Estimate the year when most oil was produced from UK oil fields.

c **i** Describe the general trend in oil production in the UK since the beginning of this century.

ii Suggest a reason for this trend.

4 The graph shows the production and consumption of natural gas in the UK between 2001 and 2011.

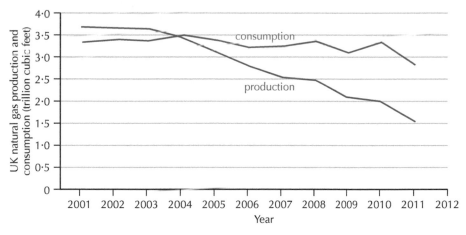

Source: US Energy Information Administration

a **i** In which year were the amounts of gas produced and consumed the same?

ii Describe the general trend in gas production between 2001 and 2011.

b Suggest how the UK gas companies managed to meet the demand for gas since consumption has been higher than UK production.

5 The table shows the main countries to which the UK exported crude oil and the quantity exported in 2015.

Country	Quantity of crude oil exported (millions of tonnes)
Netherlands	8·7
Germany	6·4
South Korea	5·3
France	2·0
Spain	2·0

Present the information in a bar chart.

6 The following is part of a report about fracking. Read it and answer the questions below.

Fracking is a process by which natural gas trapped underground in shale beds can be extracted. In 2016 the UK Government gave the go-ahead for a company to start exploring an area in Lancashire. This is the first time fracking has been allowed since a minor earthquake, thought to have been caused by fracking, stopped drilling in 2011. A mixture of water, sand and chemicals is pumped underground and forces the gas out of the shale. Campaigners say that the process contaminates drinking water and destroys the local environment. The UK Government estimates that fracking will provide us with a new energy source and create thousands of new jobs. Despite having huge shale beds in Scotland, the Scottish Government has banned fracking completely. They say that fracking undermines the country's efforts to tackle climate change. A spokesperson for the oil and gas industry claimed that without fracking, Scotland could be importing three quarters of its gas from unstable countries by 2020.

a Name the fossil fuel extracted during fracking.

b Give **three** reasons why some people are in favour of fracking.

c Give **three** reasons why some people are against fracking.

d Suggest why fracking could 'undermine the country's efforts to tackle climate change'.

e Suggest why the report could be considered as giving a balanced view of fracking.

Exercise 10B Burning fuels

This exercise includes coverage of:
N4 Fuels
CL4 Materials SCN 4-16b

1 Copy and complete the following paragraph. Use the word bank to help you.

carbon	carbon dioxide	carbon monoxide	combustion	complete
energy	incomplete	limited	oxidation	oxygen	unlimited	water

Throughout the world fossil fuels are burned to produce _ _ _ _ _ _. When a fuel burns it reacts with _ _ _ _ _ _ from the air so the process is an example of _ _ _ _ _ _ _ _ _. Burning is also known as _ _ _ _ _ _ _ _ _ _. When there is _ _ _ _ _ _ _ _ _ (lots of) oxygen present only _ _ _ _ _ _ _ _ _ _ _ _ and _ _ _ _ _ are produced. This is known as _ _ _ _ _ _ _ _ combustion. If there is _ _ _ _ _ _ _ (not a lot of) oxygen, _ _ _ _ _ _ _ _ _ _ _ _ and _ _ _ _ _ _ are formed. This is known as _ _ _ _ _ _ _ _ _ _ combustion.

2 Methane is the main hydrocarbon found in natural gas.

   a State what a **hydrocarbon** is.

   b Write a word equation for the complete combustion of methane.

   c What name is given to a reaction, such as combustion, which gives out energy?

3 Gas welding torches use a mixture of pure oxygen and a hydrocarbon called acetylene to produce a flame hot enough to cut through steel. Carbon dioxide and water are produced.

   a State what type of combustion is taking place.

   b i State the test for carbon dioxide.

     ii Describe **two** tests used to show that water is produced.

4 There are strict safety laws concerning the maintenance of gas fires and boilers due to the possible production of carbon monoxide.

   a Explain how carbon monoxide is formed during combustion.

   b i State why there is concern over the production of carbon monoxide.

     ii Suggest why it is recommended that all homes with gas appliances should be fitted with carbon monoxide detectors.

   c Write the formula for carbon monoxide.

   d Name the element often produced at the same time as carbon monoxide.

5 A group of students noticed that when they were heating a beaker of water with a Bunsen burner, a black substance formed on the outside of the beaker.

   a State what the black substance is likely to be.

   b i Suggest what type of flame the students were using.

     ii Explain how the type of flame in part **i** is produced.

   c What name is given to the type of combustion which produces the black substance?

6 Carbon monoxide is produced inside a car engine.

   a Suggest why carbon monoxide is formed inside a car engine.

   b Part of a car's exhaust system removes carbon monoxide as carbon dioxide.

     i Name the part of the exhaust system which removes the carbon monoxide.

     ii Copy and complete the word equation for the removal of carbon monoxide.

       carbon monoxide + _____ → _____ _____

7 The diagram shows processes which take place in living things.

   a Name:

     i process A

     ii gas X

     iii gas Y.

   b Energy is taken in from the surroundings during photosynthesis.

     State what type of reaction takes in energy.

   c Process A is exothermic.

     State what is meant by **exothermic**.

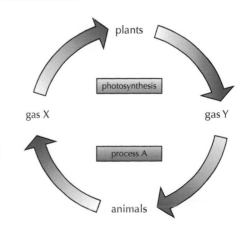

8 The fire triangle is shown.

   a Explain what the fire triangle tells us.

   b Describe how the information in the fire triangle can help us to control fires.

9 The Bunsen burner is a common piece of laboratory apparatus used for heating.

Copy and complete the following sentences by choosing the correct words in the brackets.

   a When the collar on the Bunsen burner is adjusted so that (**a lot of** / **a little**) air gets into the Bunsen burner, the flame is hot.

   b When the gas tap is switched off, the flame goes out because the (**fuel** / **heat** / **oxygen**) is removed.

10 The reactants and products involved in the complete combustion of methane (CH_4) can be shown in a word equation and by using diagrams of molecular models.

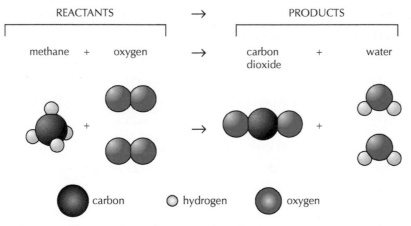

   a i State the number of atoms of each element present on the reactants side of the equation.

     ii State the number of atoms of each element present on the products side of the equation.

   b State how your answers to part **a** prove the law of conservation of mass.

11 A candle made from a mixture of hydrocarbons weighed 65 g. It was burned for 15 minutes and weighed again. It now weighed 56 g. It would seem that 9 g of the candle had 'disappeared'.

Explain what has happened to the 'missing' 9 g of candle.

11 Fuels and energy 3: The problems with fossil fuels

Exercise 11A The impact of burning fossil fuels

Hint | Make sure you can answer the questions in Exercises 6C and 6D before answering these questions.

1 Copy and complete the following passage. Use the word bank to help you.

atmosphere	burn	carbon dioxide	climate	greenhouse
heat	hydrocarbon	warming		

Fossil fuels are mainly made up of _ _ _ _ _ _ _ _ _ _ compounds. _ _ _ _ _ _ _ _ _ _ _ _ _ is one of the gases which is produced when fossil fuels _ _ _ _. Carbon dioxide is known as a _ _ _ _ _ _ _ _ _ _ gas. Increased levels of carbon dioxide is one of the main reasons that the _ _ _ _ _ _ _ _ _ _ is trapping more _ _ _ _ causing global _ _ _ _ _ _ _ which is affecting the world's _ _ _ _ _ _ _.

2 Fossil fuels are made up mainly of hydrocarbons with some impurities. One impurity is sulfur. Sulfur dioxide is produced when sulfur is burned. Sulfur dioxide is an acidic gas.

a Write a word equation for the reaction of sulfur with oxygen to produce sulfur dioxide.

b Write the formula for sulfur dioxide.

c i Name the environmental problem to which sulfur dioxide contributes.

ii Give **one** example of damage caused to the environment as a result of the production of sulfur dioxide.

3 It has been estimated that the UK has more than 200 years' worth of coal underground. One reason given by the UK Government for not using these coal reserves is the cost of getting it out of the ground.

Suggest another reason why the Government does not want to burn coal.

4 In many Chinese cities smog, a poisonous combination of smoke and fog, causes many health problems. Some local authorities are closing coal-fired power stations in an attempt to solve the problem.

a Suggest why closing coal-fired power stations might reduce the amount of smog.

b Give **one** example of how smog can affect your health.

5 The whole of Edinburgh is a smoke control area.

   a Suggest what **smoke control area** means.

   b Edinburgh was known as 'Auld Reekie'. Many of the buildings built before 1950 had a coating of black soot.

    Suggest how Edinburgh got its title 'Auld Reekie' and where the soot came from.

6 The leaders of four major cities – Paris, Mexico City, Madrid and Athens – say they intend to ban the use of diesel-powered vehicles by 2025. This is in an attempt to reduce the amount of pollution and reduce the risk of health problems caused by the exhaust emissions from diesel vehicles.

   a State the name of **two** pollutants found in the exhaust emissions from diesel engines.

   b Give **one** example of how diesel exhaust emissions can affect our health.

   c Suggest why some pedestrians and cyclists in these cities wear masks covering their nose and mouth.

7 The Scottish Government want to increase the proportion of low carbon emission buses on our roads to 50% by 2032. In 2017 the proportion was 10% and the Government's aim is to double this by 2021, reach 30% by 2025 and 40% by 2028.

   a Present the information in the paragraph above in a table. Use the headings 'Year' (starting with 2017) and 'Proportion of low carbon emission buses (%)'.

   b Give **one** reason why the Government wants to reduce the carbon emissions from buses.

8 In 1910 there were 150 glaciers in Glacier National Park in Montana, USA. The number now is 30 and they are much smaller than they were. This is thought to be due to the temperature of the Earth steadily increasing.

   a What name is given to the effect which is thought to be causing the Earth to heat up?

   b State what is causing this effect.

9 The ground in certain parts of Siberia is frozen to depths of several hundred metres all year round. Recently, however, a hole in the ground has appeared. It is called the Batagaika Crater and is nearly 1000 metres long and up to 100 metres deep. It is thought to be caused by the ground thawing. This has resulted in methane, a greenhouse gas, being released into the atmosphere.

   a Suggest what might be causing the ground to thaw.

   b i What could be the effect of releasing more methane into the atmosphere?

     ii What effect could this in turn have on the size of the crater and the amount of methane being released into the atmosphere?

12 Fuels and energy 4: Meeting energy needs in the future

Exercise 12A Sustainable sources of energy

1 Copy and complete the following paragraph using the word bank to help you.

burn	carbon dioxide	fossil	global	greenhouse
non-renewable	nuclear	renewable	solar	sustainable

Our _ _ _ _ _ _ fuels will run out eventually so we need to use other energy sources. We also need to use less fossil fuel to reduce the amount of _ _ _ _ _ _ _ _ _ _ _ _ _ produced when fossil fuels _ _ _ _. Carbon dioxide is a _ _ _ _ _ _ _ _ _ _ gas which causes _ _ _ _ _ _ warming. We need to develop energy sources which reduce carbon dioxide emissions and which are _ _ _ _ _ _ _ _ _ _ _ – they will be able to be used by future generations. They include _ _ _ _ _ _ _ _ _ energy sources such as wind and _ _ _ _ _ power and _ _ _ - _ _ _ _ _ _ _ _ _ energy sources such as _ _ _ _ _ _ _ energy.

2 Nearly two billion people in developing countries burn dried cow dung and crop waste to provide energy. Both of these fuels come from plant sources.

a What name is given to this type of energy source?

b Give the name of **one** other fuel which is obtained from a plant.

c Used vegetable oil is an example of a plant material which can be used to make fuels.

  i What name is given to fuels made from plants?

  ii Name the fuel made from used vegetable oil.

3 Ethanol can be used as a fuel. It is made from sugar cane, which is a renewable resource. Brazil is the world's biggest producer of sugar cane. Ethanol has been used as a fuel in Brazil for many years. It can be used on its own or mixed with petrol. Ethanol produces less carbon dioxide than petrol when it burns.

a State what is meant by a **renewable resource**.

b Suggest why ethanol has been used in Brazil for many years.

c State why it is important that ethanol produces less carbon dioxide than petrol when it burns.

d A fuel called E18 contains 18% ethanol and 82% petrol.

  What are the percentages of ethanol and petrol in a fuel called E25?

e Scotland used to grow a vegetable called sugar beet. A factory in Cupar, Fife, processed it into sugar for use in the food industry. It closed in the 1970s because it was cheaper to import sugar cane. There is renewed interest in growing sugar beet in Scotland and other parts of the UK, but not for use in the food industry.

  Suggest another use for the sugar produced from sugar beet.

4 Hydrogen can be used as an energy source by burning it. It can also be reacted in a fuel cell, without burning it, to make electricity.

a i State the type of energy produced when hydrogen burns.

ii Name the type of reaction which produces energy.

b When hydrogen is burned or used to make electricity it reacts with oxygen to form water.

i Write the word equation for the reaction.

ii Suggest why the reaction is considered to be environmentally friendly.

5 The table shows the sources of energy for making electricity in 2014. The pie chart shows the same information.

Energy source	Percentage (%)
coal, peat	40
natural gas, oil	26
hydro	17
nuclear	10
other renewables	7

a Match each energy source in the table with its sector of the pie chart.

b What percentage of the total is from fossil fuels?

c Suggest what could be done to decrease the amount of fossil fuel used to make electricity.

d Give **one** reason why most governments want to reduce the use of fossil fuel.

e Give **two** examples of renewable energy sources used in Scotland.

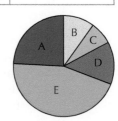

6 a State which of the following energy sources are renewable and which are finite (non-renewable).

i Coal   **ii** Waves and tides   **iii** Oil and natural gas

iv Wind   **v** Sun

b i Copy and complete the following sentence by choosing the correct word in brackets.

A sustainable resource is one which (**will** / **will not**) be available to us in the future.

ii Select the sustainable resources from the list in part **a**.

c i State **one** of the problems with using wind power and solar energy in Scotland.

ii Copy and complete the following sentence by choosing the correct word in brackets.

Solar panel sites are (**cheap** / **expensive**) to set up.

d State what type of energy is produced from the wind, waves and tides.

7 Hydroelectricity is electricity produced by water flowing from a high level to a low level. In 2015, hydroelectricity provided around 17% of the world's electricity.

a State **one** benefit of fossil fuels not being involved in the production of hydroelectricity.

b Give **two** features of Scotland's geography which make it suitable for the production of hydroelectricity.

c Copy and complete the following paragraph. Use the word bank to help you.

electricity	hydroelectric	mountains	night	released	reservoir

Cruachan power station is a pump-storage _ _ _ _ _ _ _ _ _ _ _ _ _ power station in Argyle and Bute, Scotland. It uses cheap electricity produced at _ _ _ _ _ to pump water from Loch Awe to a _ _ _ _ _ _ _ _ _ over 300 metres high in the _ _ _ _ _ _ _ _ _. The water can then be _ _ _ _ _ _ _ _ during the day to make _ _ _ _ _ _ _ _ _ _ _ when it is most needed.

13 Fuels 2: Solutions to fossil fuel problems

This chapter includes coverage of:
N4 Fuels
CL4 Planet Earth SCN 4-04b, SCN 4-05b

Exercise 13A The carbon cycle and reducing carbon dioxide emissions

> **Hint** Make sure you can answer the questions in Chapter 12 before answering these questions.

1 Copy and complete the following paragraph. Use the word bank to help you.

animals	back	carbon	carbon dioxide	cycle	energy	fossil
millions	not	photosynthesis	plants	respiration	trapped	

All living things are made from _ _ _ _ _ _ compounds. The carbon needed to make these compounds comes from _ _ _ _ _ _ _ _ _ _ _ _ _ in the atmosphere. Green _ _ _ _ _ _ remove carbon dioxide from the atmosphere by _ _ _ _ _ _ _ _ _ _ _ _ _ _. Carbon in plants is passed onto _ _ _ _ _ _ _ along food chains. Carbon dioxide passes _ _ _ _ into the atmosphere by a process called _ _ _ _ _ _ _ _ _ _ _. Respiration happens in living things and involves the production of _ _ _ _ _ and carbon dioxide. Under certain conditions carbon compounds are _ _ _ _ _ _ _ in the Earth when living things die and carbon dioxide is _ _ _ released back into the atmosphere. This is what happened _ _ _ _ _ _ _ _ of years ago when _ _ _ _ _ _ fuels were formed. These processes combined are known as the carbon _ _ _ _ _.

2 Biomass can be burned to produce energy instead of using fossil fuels.

a Give **two** examples of biomass which can be burned.

b Carbon dioxide is produced when biomass burns.

Explain why biomass can be described as carbon neutral.

c Some biomass is not suitable for burning directly but can be made into biogas. Dairy farms produce animal waste which can be used to produce biogas.

   i Give **one** example of waste biomass in the home which could be used to make biogas.

   ii Biogas contains about 60% methane and 40% carbon dioxide.

   Give **two** reasons for removing the carbon dioxide from the biogas before using it as a fuel.

d i Name **two** liquid biofuels produced from biomass which can be used as a fuel for road vehicles.

   ii Vegetable oil can be used to make a biofuel. Used vegetable oil is cheaper to use than unused.

   Suggest **one** other reason for using used vegetable oil.

3 One of the Scottish Government's climate change targets is to plant trees covering an area equivalent to about 20 000 football pitches by 2032 and to restore nearly twice that area of peat bogs.

Explain how this will help reduce the amount of greenhouse gases in the atmosphere.

4 One of the Scottish Government's aims is for Scotland to be carbon neutral by 2050. Most of the carbon dioxide emissions from industrial sites such as coal-fired power stations and cement works can be trapped by using carbon capture and storage techniques. It is estimated that, because of its oil industry, Scotland has about 35% of the carbon dioxide storage capacity in Europe.

  a Suggest what is meant by **carbon neutral**.

  b Describe how carbon dioxide can be captured and how it can be stored.

  c Suggest why Scotland has such a high capacity to store carbon dioxide.

  d Give **two** advantages of carbon capture and storage.

  e Give **two** disadvantages of carbon capture and storage.

5 The diagram shows how carbon dioxide is taken from and released back into the atmosphere.

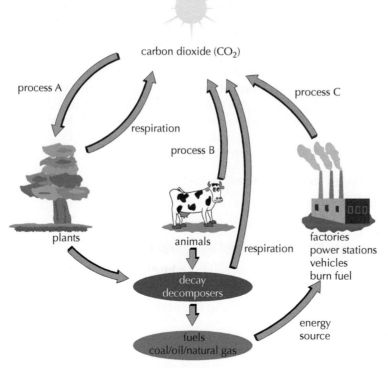

  a **i** Name processes A, B and C.

    ii What name is given to the type of fuels formed from dead plants and animals?

  b Where does the light needed for process A come from?

  c What name is given to the overall process shown in the diagram?

  d Until about 150 years ago the amount of carbon dioxide in the atmosphere was below 0·03%. It has now reached 0·04%, which is an increase of more than 30%.

    i State which human activity has led to this big increase.

    ii State why there is concern about the increasing amount of carbon dioxide in the atmosphere.

    iii Suggest **two** ways we could reduce the amount of carbon dioxide being released into the atmosphere.

6 A group of students set up an experiment to produce biogas from different biomass sources.

They heated wood shavings, then they heated hay (dried grass) in the apparatus shown. They burned the gas as it was produced.

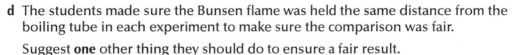

a State the aim of the experiment.

b Suggest how, by burning the gas, they could work out which biomass produced more biogas.

c One of the students suggested collecting the gas produced and comparing the volumes obtained from each biomass.

Describe how they could collect the biogas and measure the volumes produced.

d The students made sure the Bunsen flame was held the same distance from the boiling tube in each experiment to make sure the comparison was fair.

Suggest **one** other thing they should do to ensure a fair result.

7 **a** Copy and complete the following sentence by choosing the correct words in brackets.

Nuclear energy is (**unsustainable** / **sustainable**) and (**non-renewable** / **renewable**).

b The Scottish Government are interested in the use of geothermal energy.

i State the source of geothermal energy.

ii Copy and complete the following sentence by choosing the correct words in brackets.

Geothermal energy is (**unsustainable** / **sustainable**) and (**non-renewable** / **renewable**).

8 The table shows some advantages and disadvantages of different types of energy.

Advantages	Disadvantages
Can be sited offshore	Can only be built in certain areas
Can be switched on and off quickly	Dependent on weather conditions
Can be used to make biofuels	Difficult to get rid of waste
Carbon neutral	High cost
Dependable	Land and wildlife affected when reservoirs built
Energy from the Sun costs nothing	Land could be used to grow food crops
It is known in advance how much electricity will be made	Less energy is produced in the winter when it is needed the most
No greenhouse gas emissions	Needs fertilisers which can cause pollution
Plentiful source	Needs to be sited near land
Reliable	Possible greenhouse gas emissions
Very safe	Risk of accidents
	Some people think it spoils the environment

For each of these types of energy, state **two** advantages and **two** disadvantages from the table.

a Geothermal   **b** Biomass

c Wave and tidal   **d** Hydroelectricity

e Wind   **f** Solar

g Nuclear

14 Fuels 3: Hydrocarbons

This chapter includes coverage of:

N4 Hydrocarbons

CL4 Materials SCN 4-17a

Exercise 14A Fractional distillation

1 Copy and complete the following paragraph. Use the word bank to help you.

> boiling fractional fractions heated hydrocarbons
> mixtures oil refining separated

Crude _ _ _ is a complicated mixture of _ _ _ _ _ _ _ _ _ _ _ _ which is not very useful when it comes out of the ground. To make it useful, the oil is _ _ _ _ _ _ _ _ _ into smaller _ _ _ _ _ _ _ _. This is known as _ _ _ _ _ _ _ _ the oil and the first step in the process is called _ _ _ _ _ _ _ _ _ _ distillation. The oil is _ _ _ _ _ _ and compounds with similar _ _ _ _ _ _ _ points are collected together. These mixtures are known as _ _ _ _ _ _ _ _ _.

2 The diagram shows a tower which is used to separate out the mixtures in crude oil.

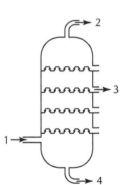

a Match the numbers on the diagram with the statements **A** to **D**.

   A Heaviest fraction out

   B Lightest fraction out

   C Preheated oil in

   D Kerosene out

b State which fraction, **A, B** or **D**:

   i has the smallest molecules

   ii contain molecules with the highest melting and boiling points

   iii is the most volatile

   iv is used as jet fuel

   v is the easiest to light

   vi burns with the smokiest flame.

c Name the process used to separate crude oil into different fractions.

3 The table shows the fractions obtained from crude oil. Copy and complete the table.

Fraction	Uses
refinery gas	
petrol	
naphtha	
kerosene	
diesel	
lubricating oil	engine oil
fuel oil	
bitumen	

4 In January 2018 an oil tanker was involved in an accident in the South China Sea. It was carrying oil known as condensate, which spilled into the sea. Condensate is a light oil which is colourless and very flammable, which made it difficult and dangerous to deal with the spillage.

   a Use the information in the paragraph above to copy and complete the following sentences by choosing the correct word in the brackets.

     i Condensate can be described as a (**volatile** / **viscous**) liquid.

     ii Condensate is likely to contain (**quite large** / **very large**) molecules.

     iii Condensate is likely produced at the (**top** / **middle** / **bottom**) of a fractionating tower.

   b State the meaning of **flammable**.

   c State the main danger involved in cleaning up the condensate.

   d Suggest why the oil's lack of colour makes it difficult to clean up.

5 The table shows the percentage of each fraction obtained from a crude oil.

Fraction	Percentage of each fraction in a crude oil (%)
refinery gas	3
petrol	13
naphtha	9
kerosene	12
diesel	14
heavy oils *	49

* For example, lubricating oils, fuel oil and bitumen

Draw a bar chart showing 'Percentage of each fraction in a crude oil (%)' against 'Fraction'.

Exercise 14B Alkanes and alkenes

1 The diagram shows a model of a methane molecule.

   a Name the elements present in methane.

   b What type of compound is methane?

   c Name the series (family) of compounds to which methane belongs.

   d Write the molecular formula for methane.

   e Draw the full structural formula for methane.

2 Copy and complete the table for the first eight alkanes.

Alkane name	Number of carbons	Number of hydrogens	Molecular formula	Full structural formula
methane	1	4		
			C_2H_6	H—C—C—H (with H's above and below each C)
	3			H—C—C—C—H (with H's above and below each C)
butane			C_4H_{10}	
		12		H—C—C—C—C—C—H (with H's above and below each C)
	6	14		
heptane			C_7H_{16}	
				H—C—C—C—C—C—C—C—C—H (with H's above and below each C)

3 a Copy and complete the table for the first seven alkenes.

Alkene name	Number of carbons	Number of hydrogens	Molecular formula	Full structural formula
ethene				C=C (with two H's on each C)
		6	C_3H_6	
butene	4	8		
				C=C—C—C—C—H (with H's on carbons)
hexene			C_6H_{12}	
				C=C—C—C—C—C—H (with H's on carbons)
	8	16		

b Explain why there is no compound called methene.

4 Copy and complete the following sentences by choosing the correct word in the brackets.

a Alkanes and alkenes are both families of (**hydrocarbons** / **carbohydrates**).

b Nonane is an (**alkane** / **alkene**).

c Decene is an (**alkane** / **alkene**).

d The carbon atoms in alkanes are joined by (**single** / **double**) bonds.

e (**Alkanes** / **alkenes**) contain a carbon-to-carbon double bond.

f (**Alkanes** / **alkenes**) decolourise bromine solution quickly.

5 Up to 50% of crude oil can be made up of heavy oils and bitumen for which there is limited use. The large molecules in these fractions can be broken down to make fractions which are more useful to us.

a Give **one** use for bitumen.

b i Name the process used to break down large molecules into smaller ones.

ii State **one** use for the fractions obtained by breaking down large molecules.

c The equation shows the products formed when a hydrocarbon molecule is broken down into smaller molecules.

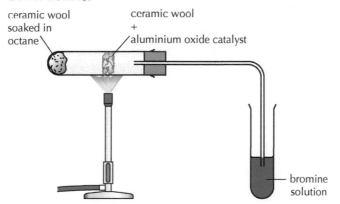

i Draw the structural formula for compound Z and name it.

ii Name the series of hydrocarbons to which Z belongs.

iii Name compounds X and Y.

iv Name the hydrocarbon series to which X and Y belong.

d The diagram shows the apparatus which can be used in the laboratory to break down octane.

i Describe what you would see happening to the orange bromine solution.

ii Explain your answer to part **i**.

iii State what a catalyst does in a reaction.

iv There is a danger in this experiment of the bromine solution being sucked back into the boiling tube when the Bunsen is switched off and an explosion occurring.

Describe what should be done to avoid this happening.

15 Everyday consumer products 1: Plants for food

This chapter includes coverage of:

N3 Everyday consumer products

CL3 Planet Earth SCN 3-02a

Exercise 15A What's in our food?

1. Copy and complete the following paragraph. Use the word bank to help you.

   | balanced | carbohydrates | compounds | diet |
   | foods | healthy | nutrients | oils | vitamins |

   Plants are _ _ _ _ _ which are a good source of the _ _ _ _ _ _ _ _ _ we need for us to grow and keep our bodies _ _ _ _ _ _ _. We need to eat the correct foods so we get the essential elements and _ _ _ _ _ _ _ _ _ our bodies need. This is known as a _ _ _ _ _ _ _ _ diet. There are five main food groups that should be part of our _ _ _ _ to provide the nutrients we need. They are _ _ _ _ _ _ _ _ _ _ _ _ _, fats and _ _ _ _, proteins, _ _ _ _ _ _ _ _ and minerals.

2. a The table shows the five main food groups we need to eat and some ways our body uses them.

   Match each food group to the correct use.

Food group	How our body uses the food group
1 Carbohydrate	A Fights infection and prevents disease
2 Proteins	B For healthy teeth, bones and blood
3 Fats and oils	C For growth and repair of body tissue
4 Vitamins	D A quick supply of energy
5 Minerals	E A concentrated supply of energy and an insulator against the cold

   b The table shows plant foods which can provide the food groups our bodies need for healthy growth.

   Match each food group to the correct plant food.

Food group	Plant foods
1 Carbohydrate	A Spinach and oatmeal
2 Proteins	B Bread and pasta
3 Fats and oils	C Oranges and kiwi fruit
4 Vitamins	D Butter and nuts
5 Minerals	E Beans and lentils

   c What word is used to describe a diet which supplies all of the food groups our bodies need?

   d Suggest why it is not healthy to eat too much food containing only carbohydrates, fats and oils.

3 The five food groups we need for a healthy body are: carbohydrates (starch and sugars); fats and oils; protein; vitamins; minerals.

Tests for these food groups were carried out on different foods. The foods tested were: potato, lean meat and glucose (a sugar).

a Potato turned iodine solution blue–black.

Name the carbohydrate present in potato.

b When heated with soda lime, lean meat turned pH paper blue.

Name the food group present in lean meat.

c Copy and complete the following sentence by choosing the correct words in the brackets.

Benedict's solution will change from (**blue / brown**) to (**black / orange**) when glucose is tested.

d Some pastry was pressed against a piece of filter paper and a greasy mark was left.

Which food group is likely to have caused the greasy mark?

4 A person's body mass index (BMI) is calculated using their weight and height. It can give an indication of how their weight affects their health.

A normal BMI is in the range 19–25. An overweight person has a BMI of 25–30. An obese person has a BMI more than 30.

a Present this information in a table with the headings 'BMI' and 'Normal / Overweight / Obese'.

b BMI can be calculated using this formula: $\text{BMI} = \dfrac{\text{weight(kg)}}{\text{height(m)} \times \text{height(m)}}$

  i Calculate the BMI of a person who is 1·2 m high and weighs 50 kg.

  ii How would a person with the BMI you calculated in part **i** be described?

  iii Which two food groups would you advise a person whose BMI was more than 25 to cut down on?

  iv Apart from reducing their intake of certain foods, what could a person do to lose weight?

  v Name **two** medical conditions that a person with a high BMI could suffer from.

5 People with anaemia have a reduced number of healthy red blood cells. This is due to a lack of iron in their diet. The diagrams show blood samples from a healthy person and a person suffering from anaemia.

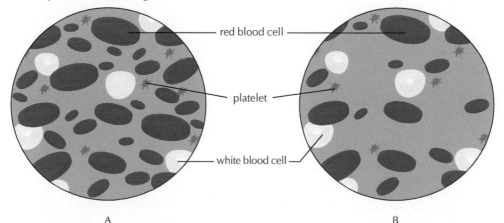

A   B

a i State which of the blood samples, A or B, is from an anaemic person.

  ii Explain your answer to part **i**.

(*continued*)

b The diagram shows the food sources of some vitamins and minerals.

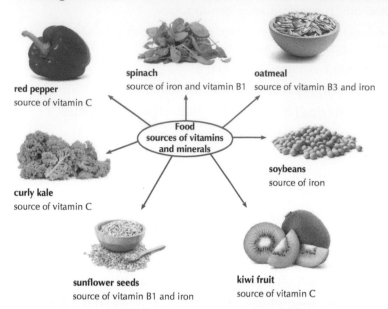

spinach
source of iron and vitamin B1

oatmeal
source of vitamin B3 and iron

red pepper
source of vitamin C

Food sources of vitamins and minerals

soybeans
source of iron

curly kale
source of vitamin C

sunflower seeds
source of vitamin B1 and iron

kiwi fruit
source of vitamin C

   i Name **one** food source which could help prevent anaemia only.

   ii Beriberi is a common disease in Africa and Asia caused by a lack of vitamin B1.
      Name **two** foods which help prevent beriberi.

   iii What other advantage is there of eating the foods you gave as answers to part **ii**?

   iv Scurvy is caused by a lack of vitamin C. Name a fruit which helps prevent scurvy.

6 Copy and complete the following sentences by choosing the correct word in the brackets.

  a i Essential fatty acids in the diet are thought to (**decrease** / **increase**) unhealthy blood fat.

    ii Essential fatty acids (**can** / **cannot**) be made in the body.

    iii The best sources of omega 3 and 6 essential fatty acids are (**plants** / **animals**).

  b i Our bodies (**can** / **cannot**) store protein.

    ii We get our protein from (**food** / **water**).

    iii There are thought to be health benefits from eating more (**plant** / **animal**) protein.

Exercise 15B Alcohols and fertilisers

This exercise includes coverage of:

N3 Everyday consumer products

1 Copy and complete the following paragraph. Use the word bank to help you.

4%	15%	40%	alcohol	distilled	heating	fermentation
plants	separates	sugars	wine			

Alcohol can be produced from the _ _ _ _ _ _ found in _ _ _ _ _ _ by a process called
_ _ _ _ _ _ _ _ _ _ _ _. Beer and _ _ _ _ are examples of drinks containing _ _ _ _ _ _ _. Beer
generally has around _ _ alcohol. Wine can contain up to _ _ _ alcohol. Drinks such as
brandy and whisky contain around _ _ _ alcohol. To increase the alcohol content of
a drink the fermented solution has to be _ _ _ _ _ _ _ _ _. This involves _ _ _ _ _ _ _ the
solution until the alcohol _ _ _ _ _ _ _ _ _ from the water and is collected.

2 To make it easier to work out how much alcohol is in a drink it is often measured in units. The table shows the number of units in various drinks.

Alcoholic drink	Alcohol content (units)
small measure (25 ml) of whisky (43%)	1·00
small glass (125 ml) of wine (14%)	1·75
one pint of beer (4%)	2·30

a The Scottish Government recommends that an adult drinks no more than 14 units of alcohol a week.

Suggest why the Government has recommended such a limit on the amount of alcohol drunk.

b Use the information in the table and a calculator to help you work out the following.

  i The number of pints of beer equal to 14 units

  ii The number of small glasses of wine equal to 14 units

  iii The number of small measures of whisky equal to 14 units

c In Scotland the drink–drive limit is around 2–3 units of alcohol.

  i Suggest why there is a legal limit on the amount of alcohol a driver is allowed to have.

  ii What is the best advice, regarding drinking alcohol, to give someone if they intend to drive?

3 Plants need certain elements for healthy growth.

a i What name is given to the elements needed by plants for healthy growth?

  ii Name the three main elements in the nutrients needed by plants.

  iii State where plants get the nutrients they need from.

  iv What can farmers and gardeners do to make sure plants get the nutrients they need?

  v State why plants are so important to us.

b The diagrams show two plants, grown from seeds which were planted at the same time.

A   B

  i Which plant, A or B, is likely to be short of the elements needed for healthy growth?

  ii Explain your answer to part **i**.

16 Everyday consumer products 2: Cosmetic products

This chapter includes coverage of:

N3 Everyday consumer products

CL3 Materials SCN 3-17b

Exercise 16A Plants, cosmetics and essential oils

1 Copy and complete the following paragraph. Use the word bank to help you.

carbohydrates	change	cocoa	cosmetics	creams	extracted
fat	looks	moisturising	oils	olive	shiny	skin	soft

Plant products such as _ _ _ _ _ _ _ _ _ _ _ _ _, fats and _ _ _ _ are used in the
_ _ _ _ _ _ _ _ industry. Cosmetics are used to _ _ _ _ _ _ the way the body _ _ _ _ _ or
smells. A carbohydrate extracted from seaweed is added to _ _ _ _ _ _ used to soften
the _ _ _ _ and hair. Cocoa butter is a _ _ _ obtained from the _ _ _ _ _ bean. It is in
many _ _ _ _ _ _ _ _ _ _ _ _ body creams. Plant oils like _ _ _ _ _ oil are included in
creams which keep the skin _ _ _ _ by preventing the loss of water. Half the weight
of a lipstick is due to castor oil _ _ _ _ _ _ _ _ _ from the castor bean. It forms a tough
_ _ _ _ _ film when it dries.

2 **a** Many cosmetics contains essential oils.

Why are essential oils useful in cosmetics?

b The following ingredients are found in a hand cream: shea butter, soybean oil, orange peel oil.

i Name **one** ingredient which contains essential oil.

ii Name **one** ingredient which can be classed as a fat.

iii Name **one** ingredient extracted by steam distillation.

iv Name **one** ingredient which is extracted by pressing part of the plant.

3 Essential oils are usually obtained from the following parts of plants: fruit, flowers, leaves, root.

a Suggest which part of the following plants the essential oils are obtained from.

i   ii   iii   iv

lavender   peppermint   ginger   lemon

b Which of the plants in part **a** are the source of essential oils most likely to be used in:

i toothpaste

ii shampoo

iii hand cream

iv muscle massage oils?

4 The diagram shows how essential oils can be extracted from some orange peel.

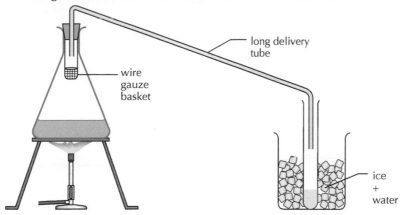

long delivery tube

wire gauze basket

ice + water

a Give the name for this method of extraction.

b Suggest what the purpose of the gauze basket is.

c Suggest why the test tube is placed in a beaker of ice.

d State **one** way you would know that an essential oil had been collected in the test tube.

e Name **two** other methods used to extract essential oils from plants.

5 Perfumes are blends of essential oils.

a i State what is meant by **blends**.

  ii State why essential oils are used to make perfumes.

b Perfumes are made up of essential oils with different notes: top, middle and base.

  i Which note is detected first when you smell a perfume?

  ii Which note can last up to 3–4 hours?

  iii Give **one** example of a base note.

  iv State why it is important to have base notes in a perfume.

6 The total spend on cosmetics in the UK in a year is about £9 billion.

The table shows the percentage of each sector of the cosmetics industry sold in the UK each year.

Sector of the cosmetics industry	Percentage sold (%)
toiletries	25
haircare	24
skincare	17
colour cosmetics	16
fragrances	

a Calculate the percentage for fragrances.

b The amount spent in each sector can be worked out using the relationship:

$$\text{amount spent in a sector} = \frac{\text{percentage}}{100} \times \text{total spend}$$

Calculate how much is spent on haircare.

17 Everyday consumer products 3: Plants for energy

Exercise 17A Carbohydrates, fats and oils

1 **a** Name the source of carbohydrates and oils such as rapeseed oil.

b Give **two** uses for fats and oils.

2 The diagram shows a model of a glucose molecule.

a Write the molecular formula for glucose.

b Work out the ratio of hydrogen atoms to oxygen atoms in a glucose molecule.

c Which group of compounds does glucose belong to?

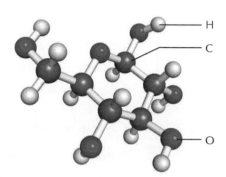

3 The diagram shows what happens when concentrated sulfuric acid is added to sucrose, a carbohydrate.

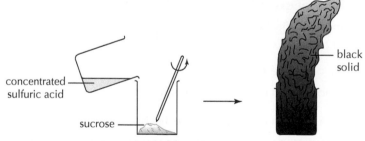

a Suggest what the black solid is.

b The sulfuric acid removes water from the sucrose. So much heat is produced during the reaction that the water turns to steam.

  i What does this suggest about which elements are present in sucrose?

  ii Name the type of reaction which produces heat.

4 **a** Copy and complete the following sentences by choosing the correct word in the brackets.

  i Glucose is a (**complex** / **simple**) carbohydrate.

  ii Starch is a (**complex** / **simple**) carbohydrate.

b Fructose is a simple sugar.

  Write the molecular formula for fructose.

c State how starch is formed in plants.

d State how plants store energy.

5 Copy and complete the following sentences by choosing the correct word in the brackets.

    a When iodine is added to starch, a (**blue–black / brown**) colour appears.

    b When a mixture of Benedict's solution and glucose is warmed, the colour changes from (**blue / orange**) to (**blue / orange**).

6 Two white powders were found in jars with no labels. It was thought that one of the powders was glucose and the other was starch.

    Describe **two** chemical tests which could be carried out to show which powder was which.

Exercise 17B Enzymes and digestion

1 Bread is a food which is high in starch. Starch gets broken down in the body into molecules which can pass through the gut wall into the bloodstream.

    a Name the process by which starch is broken down.

    b Name the molecule formed when starch is broken down.

    c Suggest why the molecules produced when starch is broken down can pass through the gut wall but starch cannot.

2 Glucose reacts with oxygen in the cells of the body. Carbon dioxide and water are produced during the reaction.

    a i Name the process taking place when glucose reacts in the cells of the body.

      ii State what else is produced during the process.

    b i Write a word equation for the reaction taking place.

      ii Complete the chemical equation for the reaction.

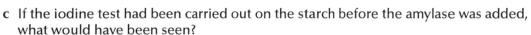

$$\underline{\hspace{2cm}} + O_2 \rightarrow \underline{\hspace{2cm}} + \underline{\hspace{2cm}}$$

3 Some students wanted to see if the enzyme amylase found in saliva was able to break down starch molecules. They set up the experiment shown in the diagram.

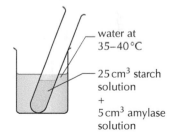

water at 35–40 °C

25 cm³ starch solution +

5 cm³ amylase solution

A sample of the mixture was tested with iodine after half an hour. The colour of the iodine did not change.

    a State the aim of the experiment.

    b What does the iodine test on the mixture after half an hour indicate?

    c If the iodine test had been carried out on the starch before the amylase was added, what would have been seen?

    d What conclusion can be made from the result of the experiment?

    e One student suggested that glucose may have been produced when the starch was broken down.

    Describe a chemical test which could be carried out to prove whether glucose was present.

    f Suggest why the temperature of the water in the beaker was 35–40 °C.

    g Enzymes are biological catalysts.

    Suggest what the enzyme does in the reaction when starch is broken down.

4 The alcohol in alcoholic drinks such as beer is produced by a process called fermentation.

  a Name the alcohol present in alcoholic drinks.

  b Name the carbohydrate which is turned into alcohol.

  c Yeast is added to a mixture of the carbohydrate and water.

   What does the yeast supply?

  d Fermentation stops when the concentration of alcohol reaches about 15%.

   i Explain why this happens.

   ii Explain how drinks with an alcohol concentration of 40% and more can be made.

  e The alcohol in alcoholic drinks is made from carbohydrates in plants.

   Explain why alcoholic drinks with different flavours can be made.

5 **a** The graph shows how well the yeast enzyme used in wine making works at different temperatures.

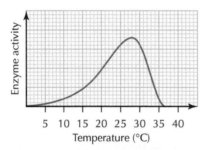

   i At what temperature does the enzyme work best?

   ii Suggest what might happen to this enzyme if the temperature is increased above 37 °C.

  b The graph shows how well two enzymes involved in the digestion of food work in the stomach.

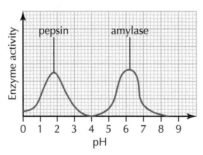

   Which enzyme works better in the most acidic conditions?

  c What term is used to describe the conditions under which an enzyme works best?

6 By law the number of units of alcohol must be shown on the label of the bottle. One unit of alcohol is equivalent to 10 cm³ of pure alcohol.

The number of units of alcohol in a bottle of any alcoholic drink can be worked out using this relationship:

$$\text{number of units} = \frac{\text{total volume of bottle (cm}^3) \times \text{percentage of alcohol}}{1000}$$

Calculate the number of units of alcohol in a 700 cm³ bottle of vodka which is 38% proof.

18 Plants to products

This chapter includes coverage of:

N3 Plants to products • **N4** Plants to products

Exercise 18A Products from plants

1 Copy and complete the following paragraph. Use the word bank to help you.

> chemists   compounds   dyes   extracting   food
> manufacture   plants   shampoos

_ _ _ _ _ _ _ _ have an important role to play in the design and _ _ _ _ _ _ _ _ _ _ _ of products which can be obtained from _ _ _ _ _ _. This includes: processing plants to produce _ _ _ _ for us to eat; _ _ _ _ _ _ _ _ _ _ compounds such as _ _ _ _ and _ _ _ _ _ _ _ _ which can be used as medicines; making new products such as soaps, _ _ _ _ _ _ _ _ and cosmetics.

2 Maize (corn) and wheat are plants which are grown and processed into important foods.

a Name a breakfast cereal made from maize.

b Wheat is ground into flour.

Give **one** example of an everyday food made from flour.

3 Coloured compounds can be extracted from certain plants. One use is as a dye for colouring textiles. Indigo, for example, is used to dye denim jeans.

a i Suggest what colour indigo is.

ii Indigo is not soluble in water.

Suggest why this is an advantage when jeans dyed with indigo are washed.

b Give **one** other use for dyes extracted from plants.

4 Plants and plant extracts are used to make many everyday products. Match the plant/plant extract to its product.

Plant/plant extract: palm oil   willow bark   wheat   saffron   argan essential oil
            aloe vera   opium poppy

Product: cosmetics   soap   breakfast cereal   aspirin   morphine
        food colouring   shampoo

5 a Copy and complete the sentences by choosing the correct word in the brackets.

i Medicines which come directly from plants are said to be (**natural** / **synthetic**).

ii Medicines made by chemists are (**natural** / **synthetic**).

b Chemists carry out research to identify compounds which can be used in medicines to treat the cause or symptom of a condition or illness.

State what term is used to describe these compounds.

c The label shows what is in a medicine used to reduce fever in young children who have a cold.

Identify the compound which treats the illness.

> **Junior Fever Relief**
>
> *Ingredients*
>
> Ibuprofen, E420 (artificial sweetener), E214 (preservative), E122 (red dye), water

(continued)

d This label appears on the box of a medicine given to children to help reduce a fever.

USES: For the reduction of fever (including post-immunisation fever) and for the relief of mild to moderate pain such as headache, sore throat, teething pain and toothache, cold and flu symptoms and minor aches and sprains.

Please read the enclosed leaflet carefully before using this product.

DOSE: For oral use.

SHAKE WELL BEFORE EACH USE.

A spoon is provided to measure doses of 2·5ml or 5ml, as required. Do not give to a child weighing less than 5kg.

Children over 7 years: Two 5ml spoonfuls to be given 3 times a day.

Children 4 to 7 years: One 5ml spoonful plus one 2·5ml spoonful (7·5ml) to be given 3 times a day.

Children 1 to 4 years: One 5ml spoonful to be given 3 times a day.

Infants 6 months to 1 year: One 2·5ml spoonful to be given 3–4 times a day.

Babies 3 to 6 months: One 2·5ml spoonful to be given 3 times a day.

Do not give to babies aged 3 to 6 months for more than 24 hours.

If your child's symptoms persist for more than 3 days, or if new symptoms occur, talk to your doctor.

Doses should usually be given every 6–8 hours, preferably with or after food.

Do not give more often than every 4 hours.

Do not exceed 4 doses in any 24-hour period.

Do not give to children under 3 months except for post-immunisation fever: one 2·5ml spoonful may be given followed by one further 2·5ml spoonful 6 hours later if necessary.

No more than 2 doses should be given in 24-hours. If fever is not reduced, talk to your doctor.

For short-term use only.

DO NOT EXCEED THE STATED DOSE

---

i Suggest why the bottle should be shaken before use.

ii What dose should be given to a child who is 10 months old?

iii What is the maximum number of doses which can be given in a 24-hour period?

iv What advice is given about what to do if the fever does not reduce?

6 Use the information in the passage to answer the questions below.

Plants have been used for medicinal purposes for thousands of years. Willow bark was used by the ancient Greeks and Egyptians to relieve fever and reduce inflammation. Willow trees are grown worldwide. In the 19th century the chemical salicin, thought to be the chemical responsible for the medicinal properties of willow, was isolated by Johann Buchner, a German chemist. Around the same time meadowsweet flowers were found to contain salicin. Meadowsweet is found in Europe and western Asia. A Scottish doctor from Dundee, Thomas MacLagan, found that salicin reduced fever and joint pain in patients with rheumatism. Further studies showed that converting salicin into acetyl salicylate (aspirin) removed some of the unwanted side effects of salicin. Aspirin is now used in preventative medicine. Aspirin has been shown to reduce the risk of heart attack in patients with high blood pressure. Aspirin thins the blood and helps reduce the risk of stroke.

a Name the active ingredient found in willow bark.

b Name **one** other source of the active ingredient and where it is found in the world.

c Give the chemical name for aspirin.

d Give **one** example of how salicin was traditionally used in the treatment of a medical condition.

e Give **one** example of how aspirin is used in modern preventative medicine.

7 **a** Codeine is a legal painkilling drug made from the opium poppy.

  i Name **one** other legal painkilling drug made from the opium poppy.

  ii Suggest why a doctor would advise that a drug such as codeine should not be taken over a long period of time.

b i Name an illegal drug made from the opium poppy.

  ii State why the drug you gave as the answer to part **i** is illegal.

c Marijuana (cannabis) is an illegal drug in the UK.

Suggest why some medical practitioners suggest that it should be given to some patients.

19 Properties of materials 1

This chapter includes coverage of:

N3 Properties of materials

Exercise 19A Materials

 1 Copy and complete the following paragraph. Choose the missing words from the word bank.

manufactured	natural	synthetic

Materials which come from plants, animals or out of the earth are said to be
_ _ _ _ _ _ _. Materials which are made from chemicals are said to be _ _ _ _ _ _ _ _ _.
Materials which are made from natural materials are said to be _ _ _ _ _ _ _ _ _ _ _ _.

2 Copy and complete the following sentences by choosing the correct word in the brackets.

a Leather is an example of a material which is (**natural** / **manufactured** / **synthetic**).

b Glass is an example of a material which is (**natural** / **manufactured** / **synthetic**).

c Plastic is an example of a material which is (**natural** / **manufactured** / **synthetic**).

3 **a** The table shows some everyday items and some materials. Match each item to the material it is made from.

Item	Material
1 takeaway hot drinks cup	**A** ceramic
2 cooking pot	**B** cotton
3 bathroom wall tile	**C** plastic
4 jeans	**D** steel

b Which of the materials in the table is natural?

4 **a** Some common materials are: ceramic, metal, plastic.

   i State which material is a good electrical conductor.

   ii State which material is the most scratch-resistant.

   iii State which **two** materials are good heat insulators.

b Which material listed in part **a** would you use to make:

   i a waterproof jacket

   ii a wash hand basin

   iii cooking foil

   iv a disposable water bottle

   v a central heating radiator

   vi heat-resistant tiles for a spacecraft?

5 Most electrical cables are made from copper wires covered with plastic. Both materials are flexible.

    a State which other property of copper makes it useful in an electrical cable.

    b State which other property of plastic makes it useful in an electrical cable.

    c Suggest why it is important that both materials are flexible.

6 A plastic soup spoon and a metal teaspoon were left in a hot cup of tea. After 5 minutes the metal spoon felt much hotter than the plastic spoon.

    a State which spoon is the better conductor of heat.

    b One student said the test wasn't fair. What should be done to make sure the comparison was fairer?

7 Copy and complete the following sentences by choosing the correct word in the brackets.

    a A plastic which can be melted and reshaped is called (**thermoplastic** / **thermosetting**).

    b A plastic which cannot be melted and reshaped is called (**thermoplastic** / **thermosetting**).

8 **a** Suggest why a thermosetting plastic is used to make an electrical socket rather than a thermoplastic.

    b A hot piece of metal was held against a plastic container and melted the plastic.

      What kind of plastic was the container made of?

    c Suggest why a thermoplastic can be recycled but a thermosetting plastic cannot be recycled.

9 Since 2014 shops in Scotland have charged 5p for a plastic carrier bag. This is to encourage us to reuse carrier bags instead of throwing them away.

    a Give **one** reason why we shouldn't throw away plastic bags.

    b Most carrier bags are made from recycled plastic. What is meant by **recycled**?

10 Novel materials have unusual properties which we can put to good use. They are sometimes called smart materials. Hydrogels can absorb hundreds of times their own weight of water. Shape memory metals can be bent out of shape but returned to their original shape when heated. Some plastics can be made to conduct electricity. Some smart materials change colour depending on the temperature.

    a Suggest why hydrogels are used in a baby's nappy.

    b Suggest why it would be useful to make the frame of a pair of glasses out of smart metal.

    c Explain why making a plastic conduct electricity is unusual.

    d Suggest why using a smart colour material in the packaging of a ready meal could be useful.

Exercise 19B Fibres and fabrics

1 **a** **i** The table lists some common fibres used to make fabrics. Copy the table and identify whether each fibre is natural or synthetic.

      ii Name the sources of the natural fibres in the table.

    b **i** State **one** property of wool which makes it useful for making scarves and gloves.

      ii Give **one** property of nylon which makes it useful for making outdoor jackets.

Fibre	Natural or synthetic
wool	
polyester	
nylon	
cotton	
silk	

(continued)

iii Suggest **one** reason why cotton is used to make towels.

iv State which property of polyester makes it good for making ropes.

v One property of silk is that it feels very soft to the touch. Suggest a use for silk which uses this property.

2 Some threads made from natural and synthetic fibres were tested using the arrangement shown. The results of the experiment are shown in the table.

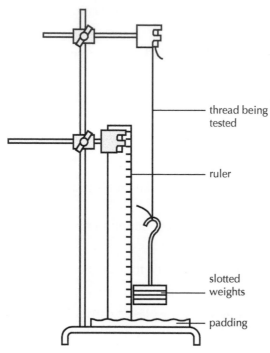

Fibre	Weight needed to break thread (g)
cotton	30
nylon	100
polyester	50
wool	10

a Suggest what the aim of the experiment was.

b i Which fibre was the strongest?

   ii Explain your answer to part **i**.

c From this experiment, which type of fibres are the weakest, natural or synthetic?

d Suggest why wool would not be a good fibre for making a fishing line.

e i The length of the fibres was kept the same in each experiment.

   Suggest why this was done.

   ii Suggest **one** other variable which should be kept the same in each experiment.

d Use the results to draw a bar chart of 'Weight needed to break thread (g)' against 'Type of fibre'.

   (Your teacher may provide you with a graph with axes and scales already drawn.)

3 The label shown gives some details about the fibres used to make a pullover.

a Draw up a table showing 'Fibre' and 'Percentage (%)'.

b Give **one** reason for mixing fibres in a material.

c The care label states that the pullover should be:

  i washed on a wool cycle (a gentle wash).

  State what this suggests about the material.

  ii ironed using a warm (not hot) setting.

  State what this suggests about the material.

  iii kept away from fire.

  State what this suggests about the material.

4 The table shows how polyester is used in a country.
The pie chart shows the same information.

Use	Percentage (%)
carpets	4
clothing	50
household textiles	13
industrial textiles	33

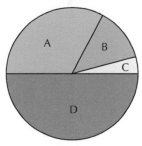

Match each use in the table to the correct area in the pie chart.

5 Copy and complete the summary. Use the word bank to help you.

breathable	heat	materials	moisture	water	windproof

Some outdoor jackets are made up of layers of different _ _ _ _ _ _ _ _ _. The
outside layer is usually _ _ _ _ _ _ _ _ _ and _ _ _ _ _ resistant. The middle layer
traps _ _ _ _ from the body to keep you warm. All the layers allow _ _ _ _ _ _ _ _
to escape from the body. This is said to make the jacket _ _ _ _ _ _ _ _ _.

6 In the 19th century Dundee was the world's biggest processor of jute. The jute was grown in India and shipped to Dundee. Jute has long fibres which can be spun into strong threads. It was used to make bags, sacks and carpet backing. In the 20th century plastic fibres such as polypropylene have replaced jute. Polypropylene is made from chemicals which come from oil. It is cheaper to make and has more useful properties than jute. Polypropylene sacks can hold a much heavier weight than jute sacks and are more hard wearing.

a Name the natural fibre mentioned in the passage.

b Name the synthetic fibre mentioned in the passage.

c State the property of the jute fibres which make it useful.

d Plastic fibres are cheaper to make than jute.

Give **one** other reason for using polypropylene instead of jute to make sacks.

7 Natural fibres have a rough texture whereas synthetic fibres are generally much smoother. Forensic scientists can use this to help identify fibres left at the scene of a crime. The photos show how threads made from synthetic and natural fibres look under a microscope.

A

B
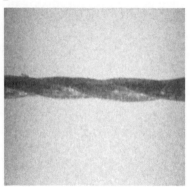

a i Which of the threads, **A** or **B**, is made from a natural fibre?

   ii Explain your answer to part **i**.

b i Suggest which of the threads is more likely to trap dust particles.

   ii Explain your answer to part **i**.

20 Properties of materials 2

Exercise 20A Polymers

1 The structure of a molecule of propene is shown.

$$CH_3 \quad H$$
$$| \qquad |$$
$$C = C$$
$$| \qquad |$$
$$H \qquad H$$

Millions of propene molecules can be made to join together to form a long-chain molecule.

a Name the long-chain molecule formed.

b What type of molecule is the long-chain molecule named in part **a**?

c Name the process by which these long-chain molecules are formed.

d What name is given to small molecules, such as propene, which join to form long-chain molecules?

2 Plastic bank notes are now used in the UK instead of paper ones.

a One advantage of plastic notes is they are not damaged if accidentally put through the wash.

   i State which property of plastic is shown by the fact that a plastic bank note is not damaged in a washing machine.

   ii Give **one** other property of plastic which makes it a better choice for a bank note than paper.

b Although it can go through the wash, a plastic bank note should not be ironed.

   i Suggest a reason for this.

   ii Plastic bank notes can be cut up and recycled.

   State what is meant by **recycled**.

   iii State which type of plastic can generally be recycled.

3 Cellulose is made in plants when millions of glucose molecules join together.

a Name the monomer in the process.

b Name the polymer.

c Name the process taking place.

d What evidence is there to indicate that the polymer is natural?

4 Copy and complete the table, which shows the names of some monomers and polymers.

Monomer	Polymer
ethene	
chloroethene	
	poly(propene)
	poly(phenylethene)

5 The properties of polymers can be changed so they can be put to different uses. Polythene (PE) is one of our most widely used polymers. Two common forms are LDPE and HDPE. LDPE is easy to stretch but not very strong. HDPE is more rigid (stiffer) than LDPE and stronger.

   a i State which form of polythene would be better for making detergent bottles.

     ii Explain your answer to part **i**.

   b i State which form of polythene would be better for making cling film.

     ii Explain your answer to part **i**.

6 Nearly all the natural rubber produced in the world comes from the rubber tree. Most of this rubber is used to make vehicle tyres. The rubber has to undergo a process called vulcanisation before it can be made into tyres.

The table compares the properties of natural rubber and vulcanised rubber.

   a Suggest **one** property of vulcanised rubber which makes it easier to handle than natural rubber during tyre manufacture.

Natural rubber	Vulcanised rubber
soft	hard
spongy	rigid (stiff)
sticky	non-stick
little resistance to rot	resistant to rot

   b Some tyres have a lifespan of up to 10 years. State **two** properties of vulcanised rubber which contribute to this long lifespan.

   c i Natural rubber is a thermosoftening polymer. State what is meant by **thermosoftening**.

     ii Vulcanised rubber is a thermosetting polymer. State what is meant by **thermosetting**.

     iii Suggest why a tyre should be made from a thermosetting polymer rather than a thermosoftening polymer.

     iv Suggest why vehicle tyres can't be recycled.

7 Most plastic carrier bags are made from **recycled** plastic. Plastics are examples of **polymers**. Only **thermosoftening** polymers can be recycled.

   a Give the meaning of the terms in **bold**.

   b Explain why only thermosoftening polymers can be recycled.

8 a A plastic hot drinks cup had this symbol stamped on the bottom of the cup.

    What does the symbol mean?

   b The cup also had PS stamped on it. This is a short form of the name of the plastic.

    Suggest what the full name of the plastic might be.

   c The cup can be recycled. What type of plastic is the cup made from?

9 The table shows the names of some common polymers and the monomers from which they are made. (Only one of the monomers used to make nylon is shown.)

Name the harmful gas or gases which can be produced when each of the plastics burn.

Plastic (common name)	Structure of monomer
polythene	H H | | C=C | | H H
PVC	Cl H | | C=C | | H H
nylon	H H | | N—(CH$_2$)$_4$—N | | H H

10 The table summarises what would be observed when various plastics are burned.

Type of plastic	Still burns when taken out of flame?	Flame colour	Drips?	Smell	Smoke?	Speed of burning
nylon	✔	blue, yellow tip	✔	burning wool	grey	slow
polystyrene	✔	yellow	✔	choking	thick black, with soot in the air	fast
polythene	✔	blue, yellow tip	✔	burning wax	black	slow
PVC	✘	yellow, blue edges	✘	choking	grey	doesn't burn out of flame

Use the information in the table to identify which plastics were being burned from these observations:

a a blue flame with yellow tip; black smoke

b choking grey smoke produced

c the plastic burns slowly and smells of burning wool.

11 The label shown appears on some items of living room furniture.

a Suggest what the label indicates.

b Give the name of **one** item of furniture which must have this label.

c Suggest why certain items of furniture must have this label.

RESISTANT

12 Disposing of plastic is an increasing problem as we use more and more plastic items.

a Landfill sites use up a lot of land and we are running out of areas to dump rubbish.

State **one** other problem with dumping plastic items in landfill sites.

b Biodegradable plastics are being used more widely.

i State what is meant by **biodegradable**.

ii Explain why refuse bags made from biodegradable plastic are more environmentally friendly than those made from non-biodegradable plastic.

c Some waste plastics are burned to get rid of them.

State **one** problem with burning plastics.

d Suggest **one** way of reducing the number of plastic items being disposed of.

Exercise 20B Ceramics and novel materials

1 Some of the properties of ceramic materials are:

high resistance to bending   high resistance to chemical attack
high wear-resistance   low electrical conductivity
low heat conductivity   very hard

Suggest which property of a ceramic would be most useful for:

a the ball and socket parts used in hip replacements

>
> **Hint** Some properties might be used once, more than once or not at all.

b the heat shield of a spacecraft

c the disc brake system of a Formula 1 sports car

d the lining of an acid storage tank

e body armour.

2 Hydrogels are often described as 'water-loving' polymers.

a Suggest why this is an unusual description for a polymer.

b Give an everyday use for hydrogels.

3 Envirobond is a polymer which attracts hydrocarbons but not water.

Suggest why this makes Envirobond useful for cleaning up oil spillages in the sea.

4 Sugru is a mouldable glue which its manufacturer claims can stick to almost anything.

Some of the properties of Sugru are:

electrical insulator   hard wearing   sets like rubber

waterproof   withstands high temperatures

Select the properties of Sugru illustrated by the following uses:

a repairing hiking boots

b repairing the rack inside a dishwasher

c repairing a laptop cable

d fixing to the corners of tables to protect children.

5 Graphene is a form of carbon which is one atom thick and its use is being developed in the electronics industry and for making lightweight batteries.

a Suggest **one** property of graphene which would make it suitable for these uses.

b A graphene filter has been developed which can separate salts from water. The salts get trapped and the water gets through.

Suggest how this could be used to help people in countries where there is a shortage of drinking water.

21 Properties of metals 1

This chapter includes coverage of:
N3 Properties of materials
CL3 Materials SCN 3-19b

Exercise 21A Metals

1 Copy and complete the following paragraph. Use the word bank to help you.

electrical electricity heat shapes strong weight wires

Metals are good conductors of _ _ _ _ and _ _ _ _ _ _ _ _ _ _ _. They are generally
_ _ _ _ _ _ and can be pressed into different _ _ _ _ _ _. They can be drawn
out to form _ _ _ _ _ which can be twisted together to form cables for
_ _ _ _ _ _ _ _ _ _ appliances and cables that are strong enough to hold the
_ _ _ _ _ _ of large structures such as the Forth Road Bridge.

2 Three metals, X, Y and Z, were heated and put into gas jars full of oxygen. The results are shown in the table.

Metal	Reaction with oxygen
X	slight glow
Y	bright flame
Z	bright glow

   a Put the metals in order of how reactive they are, starting with the most reactive.

   b Copy and complete the word equation for zinc reacting with oxygen.

   zinc + oxygen → _____ _____

3 Three metals, labelled 1, 2 and 3, were added to water. The results are shown in the table.

Metal	Reaction with water
1	bubbles produced slowly
2	bubbles produced quickly
3	no bubbles produced

   a Put the metals in order of how reactive they are, starting with the most reactive.

   b Suggest which of the metals is silver.

   c When calcium reacts with water, calcium hydroxide and hydrogen are formed.

   Write a word equation for the reaction.

4 Group 1 metals, such as potassium and sodium, are stored under oil.

   Suggest why this is done.

5 When metal items are left outside they can undergo a chemical reaction.

   a State the name given to this process.

   b Name the **two** substances the metal reacts with.

   c What special name is given to the process when iron reacts in this way?

6 Most car bodies are made of steel, a form of iron. The bodywork of a 30-year-old car had reacted so much that there were holes and it was unsafe to drive. A 500-year-old gold coin which was found in a field, however, showed no sign of a chemical reaction having taken place.

   a What does this tell you about how reactive each metal is?

   b State **one** way the bodywork of a car is treated to try to stop it reacting.

   c A car which spends most of its life near the sea is likely to be more corroded than a car which is kept inland, away from the sea.

   Suggest why this might be.

7 Iron, in the form of steel, is our most widely used metal.

  a Suggest how steel can be protected in these situations:

   i the body of a car

   ii gates and wire fencing

   iii farm tools left outside

   iv the inside of a can of food.

  b Cutlery can be protected by electroplating it.

   Explain what is meant by **electroplating**.

8 Two iron nails were put into a test tube with rust indicator. Rust indicator changes from green to blue when rust is present.

One test tube had pure water added. The other had salt water added.

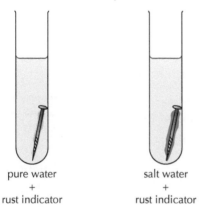

pure water
+
rust indicator

salt water
+
rust indicator

  a What does the result of the experiment indicate about the effect of salt on the rate of rusting?

  b Suggest why ocean-going ships rust more quickly than ships that sail in freshwater lakes.

9 **a** Gold is found in its pure state in the ground. Most other metals exist as compounds found in rocks known as ores. Iron, our most widely used metal, comes from haematite. Malachite is a copper-containing ore. Galena, sometimes called fools gold, contains lead. Aluminium is found in bauxite.

   Present the information in the passage about ores and their metals in a table. Use the headings 'Name of ore' and 'Metal'.

  b Copy and complete the following sentence by choosing the correct word in the brackets.

   Gold can be found in the pure state in the ground because it is very (**reactive** / **unreactive**).

  c Copy and complete the following sentences.

   i Silver can be obtained from its ore by _____ alone.

   ii Iron can be extracted from its ore by first mixing it with _____ then heating it.

   iii Aluminium is obtained by first melting the bauxite then passing _____ through it.

  d Arrange the metals in part **c** in order of how reactive they are starting with the most reactive.

Exercise 21B Metals and batteries

This exercise includes coverage of:

N3 Properties of materials

1 Copy and complete the following sentences. Use the word bank to help you.

chemicals	electricity	laptops	reaction	rechargeable	recycled

a Batteries use metals to produce _ _ _ _ _ _ _ _ _ _ _ .

b A chemical _ _ _ _ _ _ _ _ takes place inside a battery.

c Batteries stop working when the _ _ _ _ _ _ _ _ _ get used up.

d The metals in used batteries should be _ _ _ _ _ _ _ _.

e _ _ _ _ _ _ _ _ _ _ _ _ batteries can be charged up again when they run down.

f Smartphones and _ _ _ _ _ _ _ use rechargeable batteries.

2 The diagram shows how a battery can be made in the laboratory.

a State how electricity is produced.

b Suggest what would happen if one of the metals was removed from the salt solution.

c State why the bulb eventually goes out.

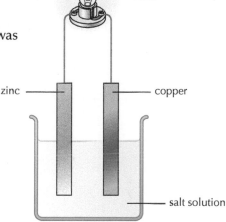

3 The diagram shows how a rechargeable battery is charged up. Bubbles of gas are seen at the lead plates.

a State what is happening during charging to indicate that a chemical reaction is taking place.

b After a few minutes the lead plates are disconnected from the power supply and connected to a bulb. The bulb lights up.

   i Suggest why the bulb lights even though the power pack is not connected.

   ii What would you see happening to the bulb after a few minutes?

c The bulb is taken out and replaced by the power pack.

   What will you see happening at the lead plates?

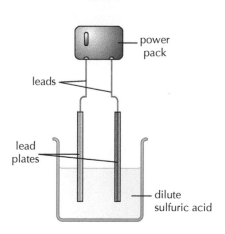

22 Properties of metals 2

This chapter includes coverage of:

N4 Metals and alloys

CL4 Materials SCN 4-16a, SCN 4-19b

CL4 Topical science SCN 4-20b

Exercise 22A Reactivity of metals and alloying

1 The table lists some terms used to describe the properties of metals.

Property	Meaning
1 ductile	**A** good at transferring heat
2 electrical conductor	**B** lightweight
3 low density	**C** easy to shape
4 malleable	**D** easily made into wires
5 thermal conductor	**E** allows electricity to pass through

Match each property to the correct meaning.

2 Hot magnesium, copper and iron were added to separate jars of oxygen.

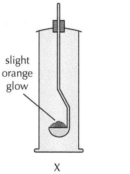
slight orange glow

X

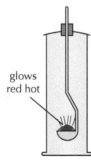
glows red hot

Y

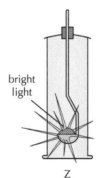
bright light

Z

a i State the names of metals X, Y and Z.

  ii Explain your choices in part **i**.

b Write a formula equation for magnesium reacting with oxygen to form magnesium oxide.

c State what kind of reaction produces energy.

3 The table summarises what happens when metals A, B and C are added to water.

Metal	Observations
A	bubbles of gas produced very quickly
B	no bubbles produced
C	bubbles of gas produced quickly

a Copy and complete the sentences by choosing the correct word in the brackets.

  i Metal A is (**calcium** / **sodium** / **tin**).

  ii Metal B is (**calcium** / **sodium** / **tin**).

  iii Metal C is (**calcium** / **sodium** / **tin**).

b The bubbles of gas produced are hydrogen. Describe a test for hydrogen.

c Lithium reacted with water to produce lithium hydroxide and hydrogen.

  i Write a word equation for the reaction.

  ii Universal indicator turned purple when added to the water after the reaction. State what this tells us about the solution formed.

4 Pieces of iron, magnesium and silver were added to test tubes containing acid.

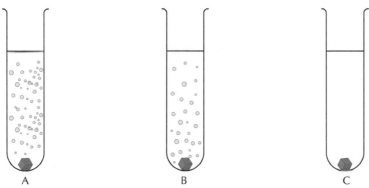

A   B   C

a i State which metal was added to each of A, B and C.

  ii Explain each of your choices in part **i**.

b Magnesium reacts with hydrochloric acid (HCl) to form magnesium chloride and hydrogen.

  Write a formula equation for the reaction.

5 Metals can be listed in order of how reactive they are.

a State the name given to the list of metals in order of how reactive they are.

b Complete the sentences by choosing the correct word in the brackets.

  i The most reactive metals are found at the (**top** / **middle** / **bottom**) of the list.

  ii The most unreactive metals are found at the (**top** / **middle** / **bottom**) of the list.

c Most of the metals are found in the ground in compounds called ores.

  i State how metals at the top of the list of metals are extracted from their ores.

  ii State how metals in the middle of the list of metals are extracted from their ores.

  iii State how metals at the bottom of the list of metals are extracted from their ores.

d Suggest where in the list of metals you would find lithium.

6 The table shows the average percentage of iron found in some ores of iron.

Iron ore	Percentage of iron (%)
magnetite	73
haematite	70
goethite	63
limonite	55
siderite	48

a Draw a bar chart of 'Iron ore' against 'Percentage of iron (%)'.

b Ores containing over 60% iron are often called usable ores.

  i Which of the ores in the table can be considered usable?

  ii Usable ores can be fed directly into a blast furnace to extract the iron. Blast furnaces operate at very high temperatures. Another substance has to be added to extract the iron.

  Suggest what the substance might be.

c i 98% of the iron produced worldwide is made into steel. Steel is a mixture of iron and other elements.

  State what name is given to a mixture of a metal and other elements.

  ii 95% of all metals used worldwide is steel.

  State the main property of steel which makes it such a useful metal.

7 The table lists the top four producers of iron ore in the world in 2015 and their percentage share of the iron ore produced. The pie chart shows the same information.

Country	Percentage share of iron ore produced (%)
Australia	47
Brazil	24
China	20
India	9

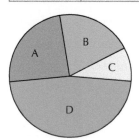

a Match each country to the correct area of the pie chart.

b China doesn't produce enough iron ore for its needs.

  Suggest how China might get the amount of iron ore it needs.

8 The timeline shows the dates that some common metals were first thought to have been extracted from their ores.

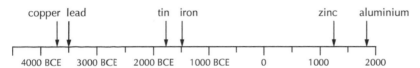

Suggest why metals such as aluminium have only recently been extracted from their ores, whereas metals such as lead and copper were extracted from their ores over 5000 years ago.

9 Brass is an example of an alloy.

a State what is meant by an **alloy**.

b The properties of some alloys are:

low melting point   resistant to corrosion   hard wearing   increased strength

Choose the property of these alloys which makes them useful for:

i solder used to join metals together in plumbing and in electrical circuits

ii coins made from cupro-nickel

iii stainless steel used to make cutlery.

10 The process of alloying metals changes the properties of the metals. The amount of gold in an item is measured in carats (ct). Pure gold is known as 24 carat gold. It is too soft for making items of jewellery such as rings so it is mixed with other metals. The less gold there is in a ring the harder it is. The table shows the composition of some of the alloys of gold.

Carats (ct)	Percentage gold (%)	Percentage other metals (%)
22	91·7	Ag: 5·0; Cu: 2·0; Zn: 1·3
18	75·0	Ag: 15·0; Cu: 10·0
14	58·3	Ag: 30·0; Cu: 11·7
9	37·5	Ag: 42·5; Cu: 20·0

a i State which gold alloy is the most hard wearing.

 ii Explain your answer to part **i**.

b i Which gold alloy would be the most expensive?

 ii Explain your answer to part **i**.

c Which type of gold is represented by the pie chart?

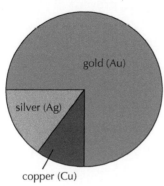

1   Three iron nails were tested under different conditions to see if corrosion took place.

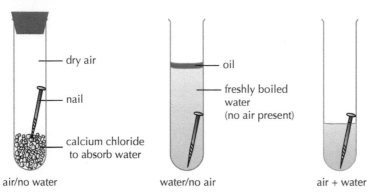

air/no water   water/no air   air + water

a i   In which test tube would you observe corrosion of the nail?

   ii Explain your answer to part **i**.

b Give the special name given to the corrosion of iron.

c When metal atoms corrode they form ions.

   i   State the charge on a metal ion.

   ii Explain how this charge occurs.

2   Here are some observations about iron and copper.

   • After a few days, brown spots form on iron railings exposed to the atmosphere.

   • The copper dome on the Usher Hall in Edinburgh starts to look green after a few years.

a i   What conclusion can be drawn about what has happened to the iron and copper?

   ii Comment on the rate at which the two metals show colour changes.

b i   Suggest what the brown spots are likely to be.

   ii Suggest what could be done to the iron railings to prevent the brown spots forming.

   iii Explain how the method you suggested in part **ii** protects the iron.

3   Steel garden tools left out in the garden will quickly show signs of rusting.

   Suggest the reason for each of the following observations.

a Steel battle tanks left in the desert since the Second World War show little signs of rust.

b Steel ships that were sunk in the seas around 100 years ago are only slightly corroded.

4 A rust indicator can be used to show if rusting is taking place.

   a Give the name of the rust indicator.

   b State the colour change which would take place in the presence of rust.

5 Strips of magnesium and copper were wrapped around different iron nails and put into test tubes with rust indicator. A third iron nail was put into a test tube with rust indicator but with no metal attached to it.

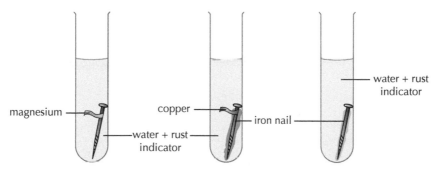

   a A slight blue colour was seen around the nail on its own.

     Suggest what this indicates.

   b There was more blue colour around the iron nail attached to the copper than around the nail on its own.

     Suggest what this indicates.

   c There was no blue colour around the nail attached to the magnesium.

     Suggest what this indicates.

   d Suggest what the aim of this experiment was.

   e State a conclusion which can be drawn from the results of this experiment.

6 Some steel items like cutlery can be protected by electroplating them.

   a Explain what is meant by **electroplating**.

   b Give **one** example of a metal used to electroplate other metals.

7 Some large steel structures such as oil rigs are protected from rusting by connecting them to the negative terminal of a d.c. power supply.

   a State what the power supply gives to the iron structure.

   b Suggest why oil rigs in the sea rust more quickly than oil rigs on land.

8 The rust which forms when iron corrodes flakes off easily and exposes fresh iron underneath.

   a What is likely to happen to the fresh iron exposed?

   b State the effect rusting has on the strength of steel.

   c Suggest why it could be dangerous if a car body was allowed to rust.

   d State **one** way that car bodywork is protected from rusting.

9 Some metals can be attached to iron to slow down rusting.

Hint Use the position of the metals in the electrochemical series to help you answer Questions 9–11.

  a What name is given to this method of protecting iron?

  b State what the metal gives to the iron to help slow down rusting.

  c i Select **one** metal from the electrochemical series which could be attached to iron in order to protect it.

   ii Give a reason for your choice of metal in part **i**.

  d Give an example of how this form of protection is used in everyday life.

10 A chemical cell can be set up using metals.

  a State what is produced in a chemical cell.

  b What name is given to solution X?

  c i Predict the direction of electron flow in the cell.

   ii Explain your answer to part **i**.

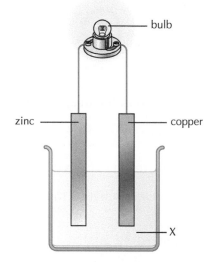

bulb

zinc

copper

X

11 Copper was connected to different metals in a chemical cell. The voltages obtained are shown in the table.

Metal connected to copper	Voltage (V)
magnesium	2·3
zinc	1·4
iron	X
lead	0·4

  a Predict the voltage produced when iron is connected to copper (X).

  b Look at the electrochemical series.

   i State which two of the four metals in the table would give the highest voltage when connected in a cell.

   ii State the direction of electron flow between the two metals you selected in part **i**.

   iii Explain your answer to part **ii**.

23 Properties of solutions

This chapter includes coverage of:

N3 Properties of materials

CL3 Materials SCN 3-16b

Exercise 23A Solubility of substances

1 Think about making a cup of coffee by dissolving instant coffee powder in water.

   a Name the solvent.

   b Name the solute.

   c Suggest how the cup of coffee could be made more concentrated.

   d State what is formed when a solid dissolves in a liquid.

   e Complete this sentence by choosing the correct word in brackets.

      A substance which dissolves in a liquid is said to be (**insoluble** / **soluble**).

2 The table shows the solubility of some compounds.

	carbonate	chloride	sulfate	nitrate
calcium	insoluble	very soluble	soluble	very soluble
iron	insoluble	very soluble	very soluble	very soluble
sodium	very soluble	very soluble	very soluble	very soluble
silver	insoluble	insoluble	soluble	very soluble

   a Which types of compounds are all very soluble?

> **Hint** Look across rows and down columns.

   b Calcium sulfate is soluble. Name another soluble compound.

   c State which metal chloride is insoluble.

   d A science technician accidentally mixed some silver carbonate powder and silver nitrate powder.

      Describe how the technician could use the solubility of each compound to help separate the mixture.

3 The table shows the amount of copper sulfate that dissolves in water at different temperatures.

Temperature of water (°C)	20	30	40
Mass of copper sulfate dissolving in 100 g of water (g)	32	38	45

Copy and complete the following sentence by choosing the correct word in the brackets.

   As the temperature (**decreases** / **increases**) the amount of copper sulfate dissolving (**decreases** / **increases**).

4 The table shows how soluble some compounds are in water.

a Which substance is the most soluble?

b Suggest why the same amount of water is used each time.

Substance	Mass dissolving in 100 g of water at 20 °C (in grams)
sodium chloride	36
calcium sulfate	0·2
copper sulfate	32
sugar (sucrose)	204

5 a Water is sometimes called the **universal solvent**. Suggest why water is given this name.

b A cookbook recommends that a certain vegetable should not be cooked in water as it loses its flavour.

   i What does this suggest about the solubility of the vegetable's flavour molecules?

   ii The cookbook recommends that the vegetable be cooked in oil so that it doesn't lose its flavour.

   Suggest why this might be.

6 The graph shows the solubility of potassium sulfate at different temperatures.

a State the solubility of potassium sulfate, in g/100 cm³ of water, at 40 °C.

b A student wanted to dissolve 16 g of potassium sulfate in 100 cm³ of water.

   State the temperature of the water needed to dissolve this amount.

c Use the graph to predict the solubility of potassium sulfate, in g/100 cm³ of water, at 80 °C.

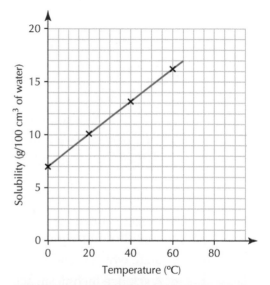

7 The table shows some solvents and their uses.

a Suggest which solvent should be used to remove a stain caused by:

   i cooking oil

   ii ink.

b A vanilla pod used to give flavour to ice cream is not very soluble in water.

   Suggest another solvent that might be used to dissolve a vanilla pod.

Solvent	What it dissolves (solute)
water	sugar, food colour, emulsion paint
alcohol (ethanol)	ballpoint pen ink, perfume, herbs and spices
acetone (propanone)	nail polish, polystyrene
white spirit	grease, oil-based paint

8 Jackets made of cotton are not very water resistant. They are sometimes coated in wax.

Suggest what this tells us about the solubility of wax in water.

9 Some stains on clothing such as grease and oil can only be removed by dry cleaning. Dry cleaning does not use water as the solvent.

a What does this suggest about grease and oil?

b The solvents used in dry cleaning are not released into the atmosphere.

   Suggest **two** reasons for not letting dry cleaning solvent escape into the atmosphere.

24 Fertilisers

This chapter includes coverage of:

N4 Fertilisers

Exercise 24A Growing healthy plants

1 All plants need nutrients.

  a State what **nutrients** are.

  b State why plants need nutrients.

  c Name the **three** nutrients needed by all plants.

  d Give **one** example of how plants growing in the wild get the nutrients they need.

  e Explain why plants are so important to animals.

2 Millions of tonnes of fertiliser are used by farmers every year.

  a Explain what a fertiliser does.

  b i Give **two** examples of natural fertilisers.

    ii For each of the examples you gave in part **i** state the source of the fertiliser.

  c More synthetic fertilisers are used than natural fertilisers.

    i Give the meaning of **synthetic**.

    ii Explain why farmers use synthetic fertilisers.

3 Synthetic fertilisers can be made by reacting acids and alkalis. Ammonium nitrate is one of our most widely used synthetic fertilisers. It is made by reacting ammonia with nitric acid.

  a Write a word equation for the reaction.

  b State what kind of reaction is taking place.

  c i Ammonium nitrate has the formula NH_4NO_3.

    Calculate the percentage composition of nitrogen in ammonium nitrate, using these RAM values: N = 14, H = 1, O = 16.

> **Hint**
>
> Percentage composition of element
>
> $$= \frac{RAM \times number\ of\ atoms\ in\ formula}{formula\ mass} \times 100$$

    ii Apart from containing nitrogen, suggest which other property ammonium nitrate has which makes it useful as a fertiliser.

  d Ammonium phosphate is also used as a fertiliser.

    i Suggest a name for the acid used to make ammonium phosphate.

    ii Suggest **one** advantage ammonium phosphate has over ammonium nitrate.

4 The table shows elements needed by plants.

Main elements	nitrogen, phosphorous, potassium, carbon, hydrogen, oxygen, sulfur, calcium, magnesium
Trace elements	iron, manganese, copper, zinc, molybdenum, boron, chlorine

a Potassium chloride (KCl) is added to fields as a source of potassium.

Calculate the percentage composition of potassium in potassium chloride, using these RAM values: K = 39, Cl = 35·5.

b Polyhalite is a naturally occurring source of potassium sulfate. It also contains magnesium and calcium sulfate.

Suggest why many farmers consider polyhalite as a more useful fertiliser than potassium chloride.

c Suggest what is meant by **trace elements**.

5 The diagram shows a bag of fertiliser.

a State what the symbols N-P-K mean.

b State what the numbers on the bag indicate.

c Suggest why a farmer might use this type of fertiliser.

6 Some plants are able to absorb the nitrogen they need from the air. This is called nitrogen fixation. Scientists are trying to develop more plants which can fix nitrogen.

Give **two** benefits of plants being able to fix nitrogen.

7 The table shows the amount of nitrogen fertiliser used worldwide in the period 1970–2010.

a Present the information in the table in a line graph.

b Extend your graph to predict the amount of nitrogen fertiliser which will be used in 2020 if it changes at the same rate.

c State the general trend in worldwide nitrogen fertiliser use.

d Suggest a reason for this trend.

Year	Amount of nitrogen fertiliser used world-wide (million tonnes)
1970	32
1980	60
1990	78
2000	82
2010	108

8 The table compares the solubility of urea ($(NH_2)_2CO$) and ammonium nitrate (NH_4NO_3) at different temperatures.

Temperature (°C)	Solubility of urea (g/100 cm³)	Solubility of ammonium nitrate (g/100 cm³)
20	110	150
40	170	300
60	250	410
80	400	580

a State the general trend in solubility of both compounds as the temperature increases.

b State why it is important for a fertiliser to be soluble.

c A major environmental issue caused by using too much fertiliser is leaching of the fertiliser.

  i State what is meant by **leaching**.

  ii Describe an environmental problem caused by fertilisers leaching into rivers and lochs.

  iii Suggest another reason why farmers should not use too much fertiliser.

d i Suggest why urea is the preferred fertiliser in parts of the world where there is annual flooding and temperatures can reach 50 °C.

  ii Suggest which fertiliser might be better in the UK where, even in the growing season, temperatures can be in single digits.

  iii Explain your answer to part **ii**.

e i The molecular formula for urea is N_2H_4CO. Calculate the percentage composition by weight of nitrogen in urea, using these RAM values: N = 14, H = 1, C = 12, O = 16.

  ii Which fertiliser, urea or ammonium nitrate, could be considered the better nitrogen fertiliser? (Hint: look back at your answer to Q3 part **c i**.)

9 The bar chart shows the amount of nitrogen fertiliser that is predicted to be needed in different parts of the world in 2018.

a Present the information given in the bar chart in a table.

b i Which area of the world uses the most nitrogen fertiliser?

  ii Suggest a reason why the area you chose in part **i** uses the most nitrogen fertiliser.

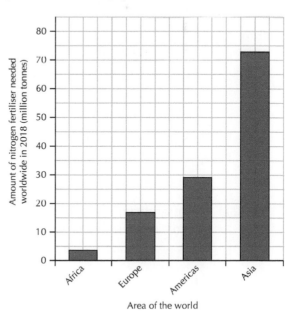

25 Nuclear chemistry

This chapter includes coverage of:

N4 Nuclear chemistry

Exercise 25A Formation of elements and background radiation

1 **a** Copy and complete the summary. Use the word bank to help you.

join	heavier	helium	hydrogen	lightest	stars

The Big Bang is thought to have produced all of the _ _ _ _ _ _ _ _ and helium in the universe. These elements are the _ _ _ _ _ _ _ _ elements. In _ _ _ _ _, hydrogen atoms _ _ _ _ with each other and form _ _ _ _ _ _. More atoms combine forming _ _ _ _ _ _ _ elements. This process continues until iron is formed.

   b Copy and complete the following summary showing what happens to hydrogen in a star.

$$\text{\_\_ hydrogen atoms} \quad \rightarrow \quad \text{one _____ atom}$$
$$H + H \quad \rightarrow \quad \text{\_\_}$$
$$\text{Atomic number:} \quad \text{\_\_} + \text{\_\_} \rightarrow \quad 2$$

   c Elements with atomic numbers greater than iron are not formed in a star because the reaction involving iron atoms fusing is too endothermic.

   State what is meant by **endothermic**.

2 **a** Copy and complete the following summaries showing element formation.

   i
$$\text{\_\_ helium atoms} \quad \rightarrow \quad \text{one _____ atom}$$
$$He + He + \text{\_\_} \quad \rightarrow \quad C$$
$$\text{Atomic number:} \quad 2 + 2 + 2 \quad \rightarrow \quad \text{\_\_}$$

   ii
$$\text{one carbon atom + one _____ atom} \quad \rightarrow \quad \text{\_\_ oxygen atom}$$
$$\text{\_\_} + \text{\_\_} \quad \rightarrow \quad O$$
$$\text{Atomic number:} \quad 6 + 2 \quad \rightarrow \quad \text{\_\_}$$

   b State the name given to the process by which atoms form new elements.

   c When atoms combine in stars energy is produced.

   Name the type of reaction in which energy is given out.

3 Copy and complete the table, which shows how elements are formed in stars. The atomic number is given in brackets.

Atoms joining	Atom formed
helium (2) and lithium (3)	_____ (5)
boron (5) and sulfur (16)	_____ (__)
oxygen (8) and fluorine (9)	_____ (__)
hydrogen (1) and _____ (__)	silicon (14)
beryllium (4) and _____ (__)	sodium (__)
magnesium (__) and _____ (__)	_____ (17)

4 Some atoms emit particles and energy. This is called radiation. Radiation is around us all the time. It comes from both natural and artificial sources.

a What name is given to the radiation which surrounds us?

b Name **two** sources of natural radiation.

c Give the main source of artificial radiation.

d Which source, natural or artificial, contributes most to the radiation which surrounds us?

5 The table shows the average amount of radiation we are exposed to from different sources. The same information is shown in the pie chart.

Radiation source	Percentage from each source (%)
breathing air	53
food and drink	11
the ground	20
from space	16

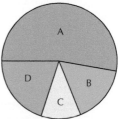

a Match each source in the table to the correct letter in the pie chart.

b State whether the sources in the table are artificial or natural.

c Suggest why there is a concern about the amount of radiation we are exposed to.

6 The graph shows how the mass of 1 g of the radioactive element technetium-99 changes with time.

a How long does it take for the 1 g sample to weigh 0·5 g?

b How much of the original 1 g sample remains after 12 hours?

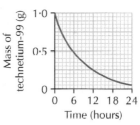

7 In April 1986 a nuclear power station in Chernobyl, Ukraine, exploded. It released radioactivity which spread over northern Europe. Some of the radioactivity was taken up by plants and animals.

a Suggest why there was concern about radioactivity getting into plants and animals.

b In 2016 a giant air-tight shield was placed over the power station.

Suggest a reason for doing this.

8 Torness nuclear power station in East Lothian can produce enough electricity to supply two million homes in the UK.

a Suggest why the people and the environment around Torness are tested for radioactivity every year.

b Name one device used to detect radioactivity.

9 The Earth's atmosphere absorbs a lot of the radiation which comes from space.

Suggest why this cosmic radiation is a bigger problem for people who travel by air a lot.

26 Chemical analysis 1

Exercise 26A Chemical hazards and analysis

1 The map shows where the major chemical industries are in Scotland.

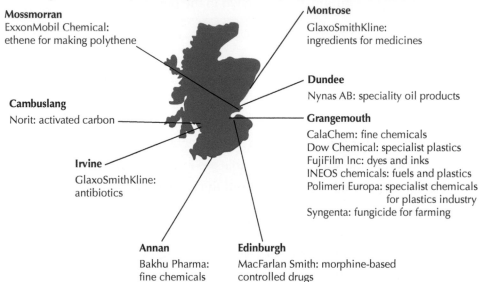

Mossmorran
ExxonMobil Chemical:
ethene for making polythene

Montrose
GlaxoSmithKline:
ingredients for medicines

Dundee
Nynas AB: speciality oil products

Cambuslang
Norit: activated carbon

Grangemouth
CalaChem: fine chemicals
Dow Chemical: specialist plastics
FujiFilm Inc: dyes and inks
INEOS chemicals: fuels and plastics
Polimeri Europa: specialist chemicals
for plastics industry
Syngenta: fungicide for farming

Irvine
GlaxoSmithKline:
antibiotics

Annan
Bakhu Pharma:
fine chemicals

Edinburgh
MacFarlan Smith: morphine-based
controlled drugs

a Name **two** places where there are companies linked to the plastics industry.

b Name **two** chemical companies which manufacture medicines and controlled drugs for medical use.

c INEOS in Grangemouth produces most of the petrol used in Scotland.

Name the fossil fuel from which we obtain petrol.

d Petrol is transported around the country in tankers.

State **one** possible hazard of transporting petrol.

2 The crude oil processed at the INEOS refinery produces a variety of chemicals. 46% of the chemicals produced is petrol and diesel. Fuel and gas oil accounts for 23% and 19% is fuel gas and jet fuel. The rest is made up of chemicals used to make plastics.

a Calculate the percentage of chemicals used to make plastics.

b Present the information given in the passage in a table. Use the headings 'Chemicals from oil' and 'Percentage produced (%)'.

3 Workers in the chemical industry have to be protected against the effects of harmful chemicals.

a State **three** safety items which should be worn by chemical workers.

b Name the piece of safety equipment which should always be worn by students when doing experiments in the laboratory.

4 People working with chemicals must know the meaning of hazard symbols on containers. The lists show some hazard symbols and meanings.

Hazard symbol	A	B	C	D	E	F
Meaning	**1** Harmful to the environment	**2** Irritant but if used properly can be used safely	**3** Catches fire easily	**4** Very toxic (poisonous)	**5** Explosive	**6** Corrosive and can burn the skin

a Match each hazard symbol to its meaning.

b Some everyday household chemicals should have hazard warnings on their containers.

Select the hazard symbol (from the table) which should be on the containers of:

i floor cleaner which could cause eye irritation

ii paint stripper which can burn the skin

iii spray paint which can catch fire

iv paint brush cleaner which is harmful to plants and animals.

c i State which **two** symbols in the table might be on a bottle of acid in a school laboratory.

ii State which symbol you would expect to see on a bottle of alcohol in a school laboratory.

d i State why it is not enough just for the tanker driver to know what they are carrying in the tanker and how hazardous it is.

ii State what is done to let the emergency services know what a tanker is carrying and how hazardous it is.

5 A factory is suspected of letting acid escape into a nearby river.

a Describe how you would test the acidity of the water.

b State how you would know from the results of your test if the water was acidic.

c Suggest why you should test samples of the river water taken before it reaches the factory waste pipe as well as samples at the waste pipe.

6 Lead and chromium are metals which can cause health problems if they are present in drinking water. Tests can be carried out on water samples to find out if lead and chromium are present. If lead is present then an orange colour is seen when potassium chromate is added to the water.

a The diagram shows how you test water for the presence of lead.

Identify X, Y and Z.

b If chromium is present in water, a light yellow colour is seen when barium chloride is added to the water.

Draw a labelled diagram showing how you would carry out the test for the presence of chromium in a water sample.

27 Chemical analysis 2

Exercise 27A The role of analytical chemists and techniques

1 Suggest why an analytical chemist might test:

a the food we eat

b household cleaning fluids

c the metals present in an ore sample

d fibres found at a crime scene

e the pH of soil

f the nutrients present in soil

g the waste water from a chemical factory

h the air at a busy road junction

i the water we drink.

2 Many companies have their own analytical chemists. The government also has agencies which have their own chemists. One is called the Food Standards Agency (FSA) and another is the Scottish Environmental Protection Agency (SEPA).

a Suggest why the government think they need their own analytical chemists.

b i If a new drink is to be sold in the shops, which agency do you think is likely to have tested it?

ii Explain your answer to part **i**.

c i If some dead fish were found in a loch, which agency do you think is likely to investigate the cause?

ii Explain your answer to part **i**.

iii Suggest **one** test that the agency may have carried out.

3 A farmer tested the pH of some soil. These are the steps she followed.

> **1** Some tap water was poured into a beaker and its pH measured.
>
> **2** The water was poured into a test tube with a sample of the soil.
>
> **3** The test tube was shaken.
>
> **4** The water was separated from the soil.
>
> **5** The pH of the water was measured.

a Describe how the soil-and-water mixture could be separated.

b Suggest why the pH of the water is measured before and after it is mixed with the soil.

c Universal indicator was used to measure the pH.

Suggest a more accurate way to measure the pH.

d The pH of the water increased from 6 to 8 after it was mixed with the soil.

i State whether the water was acidic, alkaline or neutral before it was added to the soil.

ii Suggest the colour of the indicator in the water before it was added to the soil.

iii State whether the water was acidic, alkaline or neutral after it was mixed with the soil.

iv Suggest the colour of the indicator in the water after it was added to the soil.

v Suggest what could be added to the soil to reduce the pH to 7.

4 A supermarket was selling salt called 'Supersalt' used for melting icy paths. It claimed that Supersalt was a rock salt with more than 80% salt. An analytical chemist working for the local council wanted to check their claim was true. This is what he did.

> 1 Weighed out 50 g of the rock salt into a beaker.
> 2 Added 100 cm³ of water.
> 3 Stirred the mixture.
> 4 Separated the salt solution from the rock.
> 5 Dried and weighed the pure salt.
> 6 Calculated the percentage of salt using:
>
> $$\% \text{ salt} = \frac{\text{mass of pure salt}}{\text{mass of rock salt}} \times 100$$

a State why water was added to the rock salt.

b State why the water/rock salt mixture was stirred.

c State how the rock part was separated from the salt solution.

d Describe how dry salt would be obtained.

e The pure dry salt weighed 40·3 g.

Calculate the percentage of salt in the 50 g sample.

f Comment on whether the supermarket's claim was correct or not.

5 A chemist wanted to separate a mixture of two liquids and collect them for further analysis.

a Name the separation technique she should use.

b One of the liquids has a boiling point of 100 °C and the other has a boiling point of 130 °C. The mixture is heated to 120 °C.

  i State which liquid would boil off first.

  ii Suggest a name for the substance which boils at 100 °C.

6 The table shows the characteristic flame colours produced by some metals when their compounds are heated in a Bunsen flame.

a A compound produced a red flame when heated in a flame. A student said the compound contained lithium.

Explain why this might not be true.

b A chemist heated a sample of an ore. It produced a blue-green flame.

  i State which metal might be present.

  ii Suggest why knowing a metal is present in a sample of ore is not enough for a chemist to advise a company it is worth mining for the metal.

Metal	Colour
calcium	orange-red
copper	blue-green
lithium	red
potassium	lilac
sodium	yellow
strontium	red

7 A forensic scientist ran the chromatogram shown. It shows a paint sample found at a crime scene (1) and paint taken from a suspect's car (2).

  a What conclusion can be drawn from the results of the test?

  b Explain your answer to part **a**.

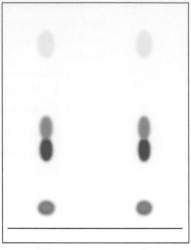

1   2

1 Crime scene paint
2 Suspect car paint

8 A food company bought a food colouring to use in the sweets they make. The label on the food colouring claimed that the dyes used in the colouring were permitted dyes. The chemists at the company used paper chromatography to separate the dyes in the bought colouring (T) and the permitted colouring (P).

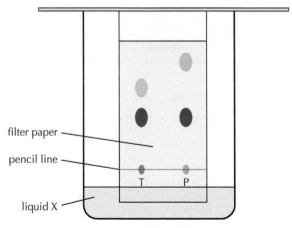

filter paper

pencil line

liquid X

T   P

  a How many dyes were in each colour?

  b i State how many permitted dyes were in the colouring bought in by the company.

  ii Explain your answer to part **i**.

  c i What advice should the chemist give to the company about whether or not they should use the food colouring?

  ii Explain your answer to part **i**.